SpringerBriefs in Environmental Science

For further volumes:
http://www.springer.com/series/8868

Daniel Priour

A Finite Element Method for Netting

Application to Fish Cages and Fishing Gear

 Springer

Daniel Priour
Ifremer
Plouzané
France

ISSN 2191-5547   ISSN 2191-5555   (electronic)
ISBN 978-94-007-6843-7   ISBN 978-94-007-6844-4   (eBook)
DOI 10.1007/978-94-007-6844-4
Springer Dordrecht Heidelberg New York London

Library of Congress Control Number: 2013938031

Printed on acid-free paper

Springer is part of Springer Science+Business Media (www.springer.com)

Contents

Chapter 1
Introduction

Abstract This chapter describes few reasons of this book: (i) the marine flexible structures such as fish cages and fishing gears could require to be studied with numerical modelling for the knowledge of their mechanics, (ii) the finite element method, well known method in engineering, is the base of the modelling used, (iii) due to the fact that these structures are mostly made of nettings and cables, these components are fully described for the finite element method, (iv) a book is a well adapted format for the description of aspects of the method.

Keywords Marine flexible structures · Finite element method

1.1 Why Fishing Cages and Fishing Gears?

Fish cages and fishing gears are generally quite large, a few tens or hundred meters, and are very flexible. Engineers and users are still trying to improve their knowledge of these flexible marine structures. This flexibility leads these structures to have different behaviours depending on the environment. The classical questions that arise are the following:

- What is the tension in the mooring line of the fish cage under certain wave, current, wind, and tide conditions?
- What is the volume reduction of the fish cage in the current?
- How dependant are the horizontal and vertical openings of the trawl on the towing speed?
- Is that the cables length of the trawl is optimal in terms of fuel consumption?

Several means are available to help engineers and users: observation at full scale, tests in flume tanks, numerical modelling. Each has its own advantages and drawbacks:

D. Priour, *A Finite Element Method for Netting*, SpringerBriefs in Environmental Science, DOI: 10.1007/978-94-007-6844-4_1, © The Author(s) 2013

- Observations give real information, but the observation area is generally very limited; it is impossible to see the whole structure at the same time.
- Tests in tanks give a lot of information, such as the behaviour of the structure in waves and in current. The main drawback is probably that the models used in tests are quite expensive, and this limits the number of tests that can be performed.
- Numerical modellings also give a lot of information, such as the tension in cables and netting twines, but they do not cover all the phenomena involved in the behaviour of the structure, such as wearing between yarns in twines or the plastic deformation of the sea bed.

1.2 Why the Finite Element Method?

Several mechanical modellings of flexible structures have been developed during the last decades. They are generally based on a decomposition of the structure into small elements in which approximations can be done. The most well-known modelling using this technique is the finite element method. This method has been widely used for mechanical modelling since the 1970s.

1.3 Why for Netting and Cable?

Nettings and cables are the main components of fishing gears and fish cages. Mechanical modellings of the structures are required to assess the behaviour of these components. Cable modellings have been described in a few publications [5, 25], but the modelling of nettings has not been given much attention. For these reasons, we attempt to fully describe a netting modelling using the finite element method. Even if the modelling of cables has been largely described, their modelling is also described here in order to propose a coherent document.

1.4 Why a Book?

Information on this finite element method for netting structures is sparse. There are portions of books and articles in journals on the topic, but there is no document that tries to group all the main matter on this subject. This book is a tentative attempt at such a publication.

Chapter 2
Finite Element Method

Abstract A brief description of the finite element method principle is proposed for the mechanics of structures. A simple example of the calculation of the perimeter of a circle is intended to highlight the principle. In case of mechanics purpose, the relevance of the concept of nodes position, of forces on the nodes and of the stiffness of the structure is described. The distinction between the local, at the scale of each finite element, and the global, at the scale of the structure, is carried out. The way, the symmetry of the structure could be taken into account in the method, is described. The boundary conditions, a fixed component such an anchor or a fixed speed such a trawler, are defined in terms of finite element method.

Keywords Description of the finite element method · Symmetry in the finite element method · Boundary conditions in the finite element method

2.1 Principle

The finite element method is a method that, at first, approximates the characteristics of a global structure by dividing it into smaller substructures called finite elements. These approximations, in the present case, are performed to estimate efforts on the vertices of these elements. These efforts depend on the position of the vertices of finite elements.

In a second step, these elements are assembled to reconstruct the overall structure and thus obtain the efforts on this structure. These efforts depend on the overall position of the vertices of the elements.

In a third step, the position of the vertices that give a zero overall effort is calculated. This position corresponds to the equilibrium position and therefore to the expected shape of the overall structure.

D. Priour, *A Finite Element Method for Netting*, SpringerBriefs in Environmental
Science, DOI: 10.1007/978-94-007-6844-4_2, © The Author(s) 2013

2.1.1 Field of Numerical Points

A field of nodes on the structure to be studied is first created. This field of numerical nodes is created so that there are many points in areas of high strain gradient. These nodes serve as the basis for creating finite elements.

The user is often in a position where he does not know *a priori* which areas are with high deformation gradients. The equilibrium positions are calculated successively, refining by adding nodes in areas with steep gradients and removing nodes in areas with low gradients.

2.1.2 Finite Elements

Finite elements are created on this field of nodes. These finite elements, in the case of our model, are of several types, depending on whether they are dedicated to cables, bars or nets.

Triangular elements are used for nets (Fig. 2.1), since the net is a surface. It seems easier to use the simplest surface, namely, the triangle. The curvature of the net can

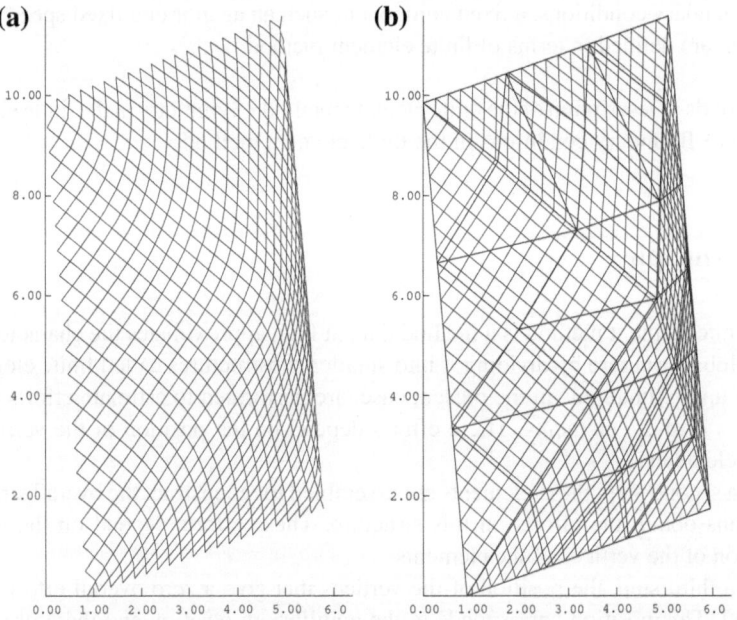

Fig. 2.1 The diamond mesh netting (**a**) is decomposed into triangular elements (**b**). The approximation in each *triangle* is that twines are parallel and therefore have the same deformation, and that the twines are elastic (Chap. 4 p. 27)

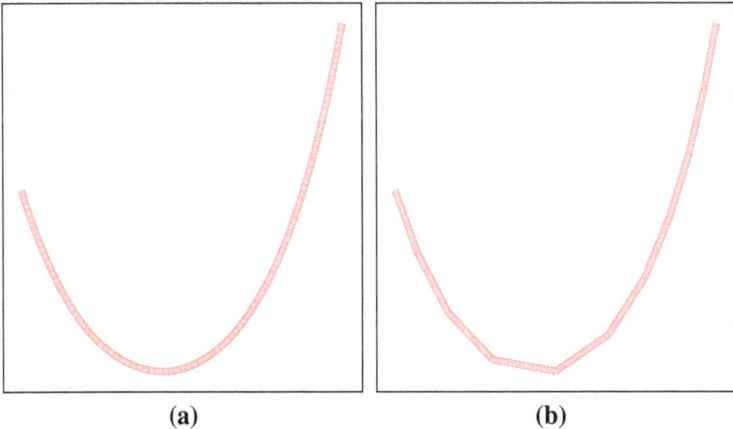

Fig. 2.2 The cable (**a**) is decomposed into bars elements (**b**). The approximation in each bar is that bars are straight and elastic (Chap. 5 p. 71)

be represented using several triangular elements. Bar elements are used for cables (Fig. 2.2).

2.2 A Simple Example

The following simple example shows the principle of splitting a global structure into several finite elements. A circle with a diameter of 1m has a perimeter of π ($2\pi R$). To assess this perimeter by the finite element approach, the circle is divided into n identical parts (Fig. 2.3). The perimeter is the sum of the length of each circle arc. The length of the arc can be approximated by the circle cord. Each cord has a length of $2Rsin(\frac{\alpha}{2})$.

The perimeter of the circle can be assessed by n times each cord length. Figure 2.4 shows the evaluation accuracy of the perimeter in function of the number of sectors for the approximation. The larger the number of elements, the greater the accuracy.

In other words, a parameter (here the perimeter) can be assessed by dividing the problem into finite elements (sectors) to be able to make acceptable approximations (the arc length approximated by the cord length). The parameter is finally assessed by rebuilding all the finite elements (sum of cord lengths). The principle of the finite element method is to discretize a structure in small (finite) elements to make acceptable approximations in each element and rebuild all the finite elements for assessing parameters on the structure.

Fig. 2.3 Polygon of n cords
inside the circle. The length of
each cord is 2 R sin(α/2). The
circle perimeter is assessed by
n times each cord length

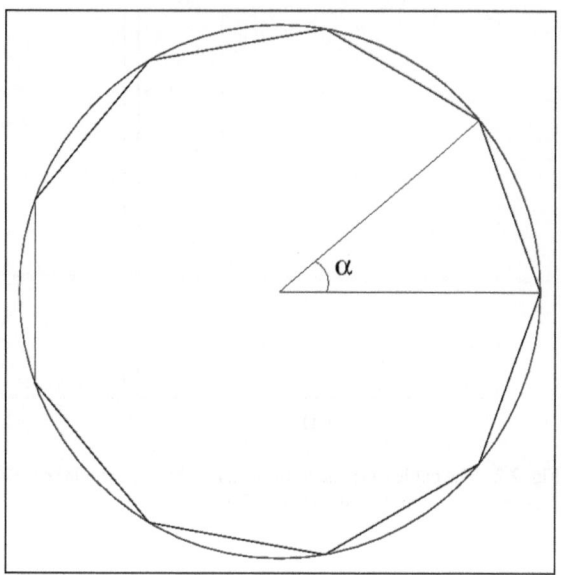

2.3 Nodes Position, Forces on Nodes, and Stiffness Matrix

In case the relationship between efforts on nodes (vertices of the elements) and their
position is established, $\mathbf{F(X)}$ is known:

$\mathbf{F}$: force on the nodes (N),

$\mathbf{X}$: node position (m).

The objective of the method is to estimate the equilibrium position ($\mathbf{X}_{final}$), that
is to say, such that

$\mathbf{F(X}_{final}) = 0$

The Newton-Raphson method is generally used to obtain this position ($\mathbf{X}_{final}$)
from an initial unbalanced position ($\mathbf{X}_{initial}$). This method iteratively calculates
the position at equilibrium. This method relies on the definition of the following
derivative:

$$F'(\mathbf{X}) = \frac{\mathbf{F(X+h) - F(X)}}{\mathbf{h}} \quad\quad (2.1)$$

F': derived efforts with respect to position (N/m),

$\mathbf{h}$: nodes displacement (m) which tends to 0.

The displacement $\mathbf{h}$ is sought if $\mathbf{X}$ is not the equilibrium position and such that
$\mathbf{X + h}$ is in equilibrium. Under these conditions:

$\mathbf{F(X + h)} = 0$

The previous equation of the derivative gives

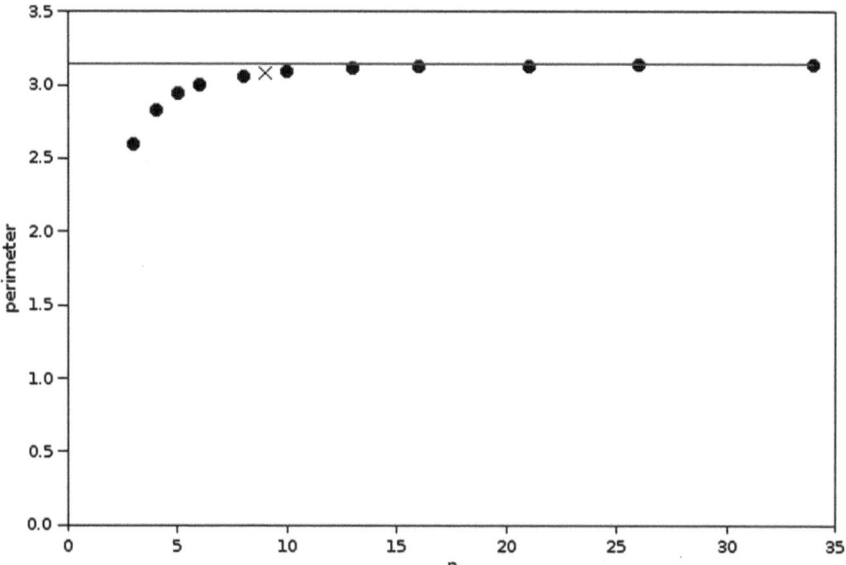

Fig. 2.4 Perimeter of the polygon (dots) in function of the number of cords (n) compared with the perimeter of the circle (line). The cross corresponds to the cords in Fig. 2.3

$$\mathbf{h} = \frac{F(\mathbf{X})}{-F'(\mathbf{X})} \tag{2.2}$$

The term $-F'(\mathbf{X})$ is called the stiffness matrix of the structure. Obviously $\mathbf{h}$ can be large, which means that the definition of the derivative is not completely respected. An iterative calculation is required:

$$\mathbf{X}_{k+1} = \mathbf{X}_k + \frac{F(\mathbf{X}_k)}{-F'(\mathbf{X}_k)} \tag{2.3}$$

k: iteration.

Starting from a position $\mathbf{X}_k$, $F(\mathbf{X}_k)$ and $-F'(\mathbf{X}_k)$ are calculated, then the displacement $\mathbf{h}_k$ is deduced and then the next position $\mathbf{X}_{k+1}$. The iterative calculation is stopped when convergence is achieved, for example when the force $F(\mathbf{X}_k)$ converges to $\mathbf{0}$.

2.4 Local and Global Forces and Stiffness

In the Chaps. 4, 5, and 6 the forces and the stiffness are described in local terms.

As mentioned earlier, the structure is split into finite elements in which forces and stiffness are calculated locally. That gives local forces $\mathbf{f}$ and local stiffness k. For example in case of element involving four coordinates, they are as in following:

$$\mathbf{f} = \begin{pmatrix} a \\ b \\ c \\ d \end{pmatrix} \tag{2.4}$$

$$k = \begin{pmatrix} e & f & g & h \\ i & j & k & l \\ m & n & o & p \\ q & r & s & t \end{pmatrix} \tag{2.5}$$

To reassemble the finite elements in the global structure, the local forces and the local stiffness have to be added to the global ones $(\mathbf{F}, K)$.

For example, if $\mathbf{f}$ and k define the force and the stiffness on an element that involves node components 3, 4, 7, and 8, taking this element into account in the global structure would mean that the local force $\mathbf{f}$ and stiffness k have to be added to the global force $\mathbf{F}$ and stiffness K, as in the following:

$$\mathbf{F}(3) = \mathbf{F}(3) + a \tag{2.6}$$
$$\mathbf{F}(4) = \mathbf{F}(4) + b \tag{2.7}$$
$$\mathbf{F}(7) = \mathbf{F}(7) + c \tag{2.8}$$
$$\mathbf{F}(8) = \mathbf{F}(8) + d \tag{2.9}$$

$$K(3, 3) = K(3, 3) + e \tag{2.10}$$
$$K(3, 4) = K(3, 4) + f \tag{2.11}$$

$$\vdots$$

$$K(4, 3) = K(4, 3) + i \tag{2.12}$$

$$\vdots$$

$$K(8, 8) = K(8, 8) + t \tag{2.13}$$

In other words:

$$
\mathbf{F} = \begin{pmatrix} \cdot \\ \cdot \\ \cdot + a \\ \cdot + b \\ \cdot \\ \cdot \\ \cdot \\ \cdot + c \\ \cdot + d \\ \cdot \\ \cdot \end{pmatrix} \tag{2.14}
$$

$$
K = \begin{pmatrix} \cdot \cdot & \cdot & \cdot & \cdot \cdot & \cdot & \cdot & \cdot \cdot \\ \cdot \cdot & \cdot & \cdot & \cdot \cdot & \cdot & \cdot & \cdot \cdot \\ \cdot \cdot \cdot + e & \cdot + f & \cdot \cdot \cdot + g & \cdot + h & \cdot \cdot \\ \cdot \cdot \cdot + i & \cdot + j & \cdot \cdot \cdot + k & \cdot + l & \cdot \cdot \\ \cdot \cdot & \cdot & \cdot & \cdot \cdot & \cdot & \cdot & \cdot \cdot \\ \cdot \cdot & \cdot & \cdot & \cdot \cdot & \cdot & \cdot & \cdot \cdot \\ \cdot \cdot \cdot + m & \cdot + n & \cdot \cdot \cdot + o & \cdot + p & \cdot \cdot \\ \cdot \cdot \cdot + q & \cdot + r & \cdot \cdot \cdot + s & \cdot + t & \cdot \cdot \\ \cdot \cdot & \cdot & \cdot & \cdot \cdot & \cdot & \cdot & \cdot \cdot \\ \cdot \cdot & \cdot & \cdot & \cdot \cdot & \cdot & \cdot & \cdot \cdot \end{pmatrix} \tag{2.15}
$$

2.5 Symmetry

In the case of symmetrical structures in a symmetrical environment it could be advantageous to use this symmetry to reduce the node number and therefore the computation times.

Figure 2.5 shows a simple bar with a symmetry plane. The plane of symmetry is OYZ and only the node of components a, b, and c, is on the plane of symmetry.

The calculation of force vector on the bar P regardless of the symmetry will give a force such as (cf. Fig. 2.5):

$$
\mathbf{F} = \begin{vmatrix} F_a \\ F_b \\ F_c \\ F_d \\ F_e \\ F_f \end{vmatrix} \tag{2.16}
$$

The stiffness matrix would be:

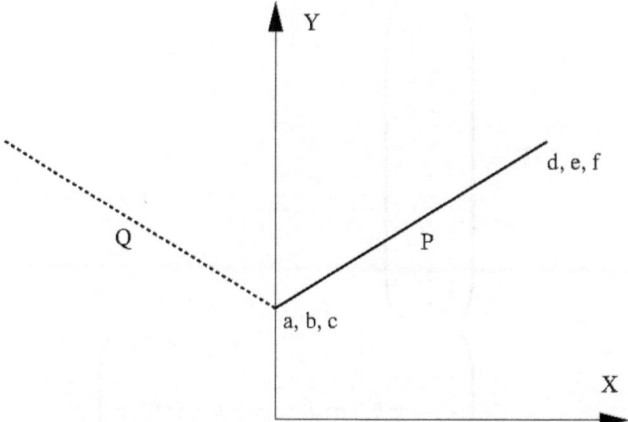

Fig. 2.5 The bar P has a node (a, b, c) on the symmetry plane. The other node (d, e, f) is outside the symmetry plane. The symmetric bar is Q

$$K = \begin{vmatrix} K_{aa} & K_{ab} & K_{ac} & K_{ad} & K_{ae} & K_{af} \\ K_{ba} & K_{bb} & K_{bc} & K_{bd} & K_{be} & K_{bf} \\ K_{ca} & K_{cb} & K_{cc} & K_{cd} & K_{ce} & K_{cf} \\ K_{da} & K_{db} & K_{dc} & K_{dd} & K_{de} & K_{df} \\ K_{ea} & K_{eb} & K_{ec} & K_{ed} & K_{ee} & K_{ef} \\ K_{fa} & K_{fb} & K_{fc} & K_{fd} & K_{fe} & K_{ff} \end{vmatrix} \tag{2.17}$$

In this case the ranking of the node coordinates is a, b, c, d, e, f.

The calculation of the total force vector on the bar taking into account the symmetry will give a force such as:

$$\mathbf{F} = \begin{vmatrix} F_a - F_a \\ F_b + F_b \\ F_c + F_c \\ F_d + 0 \\ F_e + 0 \\ F_f + 0 \end{vmatrix} \tag{2.18}$$

The stiffness matrix would be:

$$K = \begin{vmatrix} K_{aa} + K_{aa} & K_{ab} - K_{ab} & K_{ac} - K_{ac} & K_{ad} & K_{ae} & K_{af} \\ K_{ba} - K_{ba} & K_{bb} + K_{bb} & K_{bc} + K_{bc} & K_{bd} & K_{be} & K_{bf} \\ K_{ca} - K_{ca} & K_{cb} + K_{cb} & K_{cc} + K_{cc} & K_{cd} & K_{ce} & K_{cf} \\ K_{da} & K_{db} & K_{dc} & K_{dd} & K_{de} & K_{df} \\ K_{ea} & K_{eb} & K_{ec} & K_{ed} & K_{ee} & K_{ef} \\ K_{fa} & K_{fb} & K_{fc} & K_{fd} & K_{fe} & K_{ff} \end{vmatrix} \tag{2.19}$$

That gives for a symmetry plane OXY passing by the node of coordinates a, b, c:

$$F = \begin{vmatrix} 0 \\ 2.F_b \\ 2.F_c \\ F_d \\ F_e \\ F_f \end{vmatrix} \qquad (2.20)$$

$$K = \begin{vmatrix} 2.K_{aa} & 0 & 0 & K_{ad} & K_{ae} & K_{af} \\ 0 & 2.K_{bb} & 2.K_{bc} & K_{bd} & K_{be} & K_{bf} \\ 0 & 2.K_{cb} & 2.K_{cc} & K_{cd} & K_{ce} & K_{cf} \\ K_{da} & K_{db} & K_{dc} & K_{dd} & K_{de} & K_{df} \\ K_{ea} & K_{eb} & K_{ec} & K_{ed} & K_{ee} & K_{ef} \\ K_{fa} & K_{fb} & K_{fc} & K_{fd} & K_{fe} & K_{ff} \end{vmatrix} \qquad (2.21)$$

That gives for a symmetry plane OYZ passing by the node of coordinates a, b, c:

$$F = \begin{vmatrix} 2.F_a \\ 0 \\ 2.F_c \\ F_d \\ F_e \\ F_f \end{vmatrix} \qquad (2.22)$$

$$K = \begin{vmatrix} 2.K_{aa} & 0 & 2.K_{ac} & K_{ad} & K_{ae} & K_{af} \\ 0 & 2.K_{bb} & 0 & K_{bd} & K_{be} & K_{bf} \\ 2.K_{bc} & 0 & 2.K_{cc} & K_{cd} & K_{ce} & K_{cf} \\ K_{da} & K_{db} & K_{dc} & K_{dd} & K_{de} & K_{df} \\ K_{ea} & K_{eb} & K_{ec} & K_{ed} & K_{ee} & K_{ef} \\ K_{fa} & K_{fb} & K_{fc} & K_{fd} & K_{fe} & K_{ff} \end{vmatrix} \qquad (2.23)$$

That gives for a symmetry plane OZX passing by the node of coordinates a, b, c:

$$F = \begin{vmatrix} 2.F_a \\ 2.F_b \\ 0 \\ F_d \\ F_e \\ F_f \end{vmatrix} \qquad (2.24)$$

$$K = \begin{vmatrix} 2.K_{aa} & 2.K_{ab} & 0 & K_{ad} & K_{ae} & K_{af} \\ 2.K_{ba} & 2.K_{bb} & 0 & K_{bd} & K_{be} & K_{bf} \\ 0 & 0 & 2.K_{cc} & K_{cd} & K_{ce} & K_{cf} \\ K_{da} & K_{db} & K_{dc} & K_{dd} & K_{de} & K_{df} \\ K_{ea} & K_{eb} & K_{ec} & K_{ed} & K_{ee} & K_{ef} \\ K_{fa} & K_{fb} & K_{fc} & K_{fd} & K_{fe} & K_{ff} \end{vmatrix} \qquad (2.25)$$

2.6 Boundary Conditions

There are two kinds of boundary conditions: the mechanical and the geometric.

The mechanical boundary conditions are defined through forces on the structure. Such boundary conditions could be the effect of the sea bed; for example, a mooring chain lands on the bottom. This specific case is described in Sect. 6.2 (p. 87).

The geometric boundary conditions consist here in displacement boundary conditions; for example, an anchor in the sea bed could be taken into account by a null displacement, or a boat towing a gear could be defined with a null displacement in moving water. These geometric conditions are actually the conditions discussed in this section.

A null displacement for node coordinate c could be taken into account by modifying the force and the stiffness matrix. Generally speaking, the force and the matrix stiffness are such as:

$$\mathbf{F} = \begin{vmatrix} F_a \\ F_b \\ F_c \\ F_d \\ F_e \\ F_f \end{vmatrix} \qquad (2.26)$$

$$K = \begin{vmatrix} K_{aa} & K_{ab} & K_{ac} & K_{ad} & K_{ae} & K_{af} \\ K_{ba} & K_{bb} & K_{bc} & K_{bd} & K_{be} & K_{bf} \\ K_{ca} & K_{cb} & K_{cc} & K_{cd} & K_{ce} & K_{cf} \\ K_{da} & K_{db} & K_{dc} & K_{dd} & K_{de} & K_{df} \\ K_{ea} & K_{eb} & K_{ec} & K_{ed} & K_{ee} & K_{ef} \\ K_{fa} & K_{fb} & K_{fc} & K_{fd} & K_{fe} & K_{ff} \end{vmatrix} \qquad (2.27)$$

When the null displacement for node coordinate c is taken into account, the force and the stiffness matrix become:

$$\mathbf{F} = \begin{vmatrix} F_a \\ F_b \\ 0 \\ F_d \\ F_e \\ F_f \end{vmatrix} \tag{2.28}$$

$$K = \begin{vmatrix} K_{aa} & K_{ab} & 0 & K_{ad} & K_{ae} & K_{af} \\ K_{ba} & K_{bb} & 0 & K_{bd} & K_{be} & K_{bf} \\ 0 & 0 & 1 & 0 & 0 & 0 \\ K_{da} & K_{db} & 0 & K_{dd} & K_{de} & K_{df} \\ K_{ea} & K_{eb} & 0 & K_{ed} & K_{ee} & K_{ef} \\ K_{fa} & K_{fb} & 0 & K_{fd} & K_{fe} & K_{ff} \end{vmatrix} \tag{2.29}$$

These modifications of force and stiffness matrix ensure that the displacement of coordinate c is null.

Chapter 3
Equilibrium Calculation

Abstract The modelling of structure mechanics is a matter for finding equilibrium of the structure. The Newton-Raphson method for equilibrium calculation is described. This method is based on the nodes position, the forces on nodes, and the stiffness matrix. Other methods of equilibrium calculation, the methods of Newmark and of the energy minimisation, are described.

Keywords Equilibrium calculation · Newton-Raphson method · Newmark method · Energy minimisation method

3.1 Newton-Raphson Method

Finite element methods generally use the Newton-Raphson method [4] for the calculation of the equilibrium position of a mechanical structure. The equilibrium position corresponds to that position of the structure in which the sum of forces equals 0. In what follows a few simple examples are given to explain the method under three cases: one dimension, two dimensions and several dimensions.

3.1.1 One Dimension

A spring (Fig. 3.1) equilibrium is reached when the weight is equilibrated by the spring force. At this position the sum of forces equals 0. This position can be calculated using the Newton-Raphson method. In this example there is just one dimension: the vertical position (x) of the mass relatively to the spring fixation which also equals the length of the spring.

The spring equilibrium is calculated by writing the force on the mass: the weight is $-Mg$ (N), and the force of the spring is $+K\frac{x-l_0}{l_0}$ (N).

D. Priour, *A Finite Element Method for Netting*, SpringerBriefs in Environmental
Science, DOI: 10.1007/978-94-007-6844-4_3, © The Author(s) 2013

Fig. 3.1 The equilibrium of
the spring is due to the mass
weight and the spring force

With

M: mass (kg),
g: acceleration of gravity (m/s^2),
K: spring stiffness (N),
x: position of the mass along the spring axis relative to the fixed point of the spring (m),
x: length of the stretched spring (m).

In this example the stiffness is not constant in order to give a clearer explanation of
the Newton-Raphson method. K is equals to Ax. That means that longer the spring
is, the stiffer it is.

The sum of forces on the mass (curve on Fig. 3.2) is

$$F(x) = K\frac{x - l_0}{l_0} - Mg \tag{3.1}$$

or, following the previous relations,

$$F(x) = Ax\frac{x - l_0}{l_0} - Mg \tag{3.2}$$

Obviously at the equilibrium $F(x) = 0$. It is clear that this simple equation has
an analytical solution, which is

$$x = \frac{\sqrt{l_0 A \ (4gM + l_0 A)} + l_0 A}{2A} \tag{3.3}$$

The Newton-Raphson method could be used to find the length of the spring (x)
at the equilibrium. This method requires knowing the force and the derivative of the
force relatively to the position.

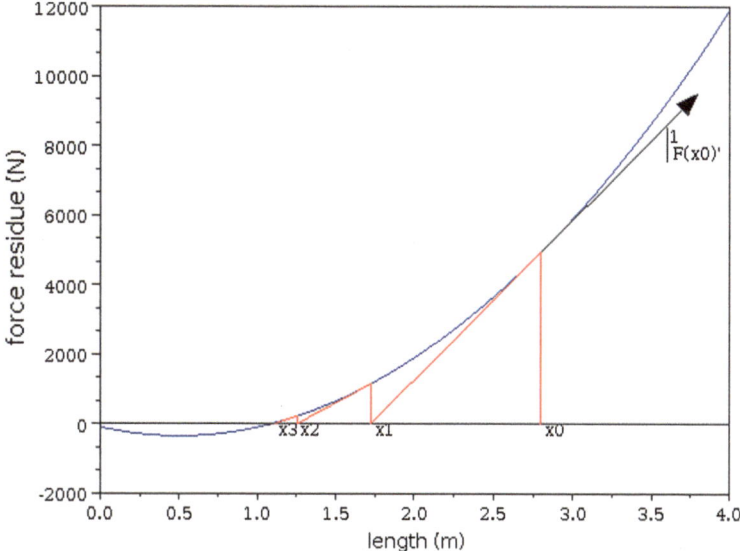

Fig. 3.2 Sum of forces on the mass function of spring length. Three Newton-Raphson iterations starting at $x = 2.8$ m are displayed. The vector tangent at x_0 is shown

The method is iterative and approximates the force curve by its tangent (shown in Fig. 3.2). From a position (x_k), the force $(F(x_k))$ and the derivative of force $(F'(x_k))$ are calculated, and a new position (x_{k+1}) can be found. This new position is generally closer to the equilibrium and is calculated as follows:

$$x_{k+1} = x_k + \frac{F(x_k)}{-F'(x_k)} \tag{3.4}$$

Figure. 3.2 shows three iterations with an initial value x_0 of the spring length of 2.8 m.

With:

The stiffness $A = 1000\,\text{N/m}$,

The mass $M = 10\,\text{kg}$,

The acceleration of gravity $g = 9.81\,\text{m/s}^2$,

The unstretched length of the spring $l_0 = 1$ m.

The stretched length at the equilibrium is 1.09 m. That means that the spring stretches 9 %.

After five iterations the equilibrium is reached or more exactly $|F(x)| < 0.1N$. The Fig. 3.2 shows 3 iterations along the curve of force. Figure 3.3 represents the reduction of the force residue ($|F(x)|$) with the five iterations.

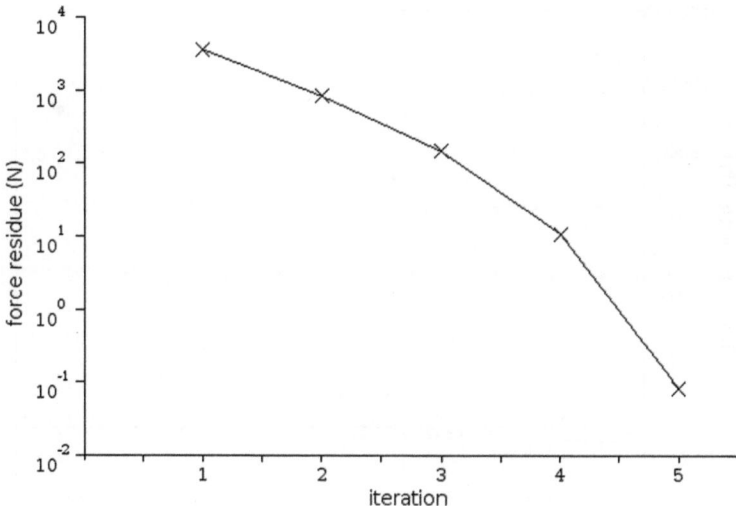

Fig. 3.3 Residue of force for each Newton-Raphson method iteration

3.1.2 Two Dimensions

In this section a simple example in two dimensions is given (Fig. 3.4): a spring
with two degrees of freedom, i.e., the horizontal (x) and the vertical (y) positions
of the mass relative to the spring fixation. The equilibrium of the system is due to
the position of the mass along the vertical and the horizontal. Figure 3.5 shows the
variation of the norm of the residue of force $\left(\sqrt{F_x^2 + F_y^2}\right)$ on the mass due to the
positions along x and y of the mass. The equilibrium point is noted by the largest
dot.

The stiffness (K) of the spring is not constant: K is equal to Al. That means that
the longer the spring is, the stiffer it is. In this condition the horizontal and vertical
forces on the mass are due to the spring length and the weight of the mass:

$$F_x = T\frac{x}{l} \tag{3.5}$$

$$F_y = T\frac{y}{l} - Mg \tag{3.6}$$

With:

$$T = Al\frac{l - l_0}{l_0} \tag{3.7}$$

$$l = \sqrt{x^2 + y^2} \tag{3.8}$$

In this case the derivative of the forces is calculated relatively to x and y:

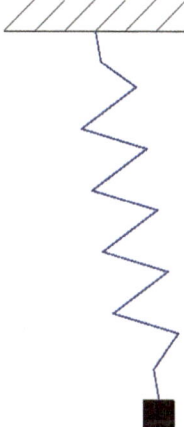

Fig. 3.4 Spring with two degrees of freedom: the *vertical* and *horizontal* positions of the mass. The equilibrium is due to the mass weight and the spring force

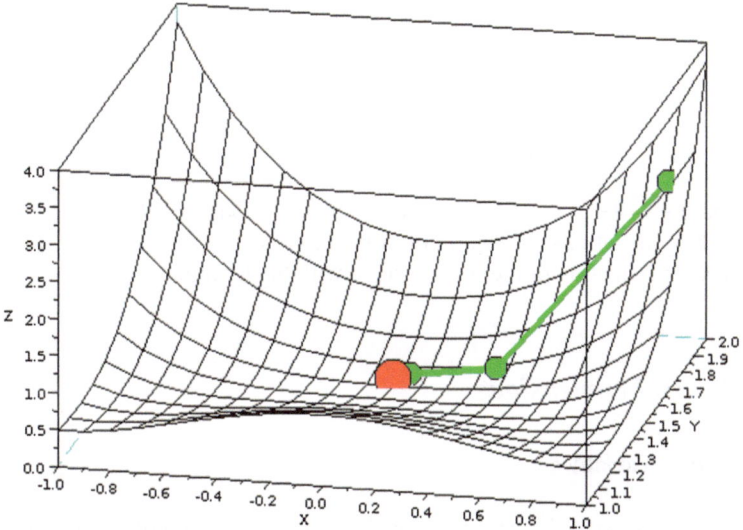

Fig. 3.5 Norm of the force ($Z = \sqrt{F_x^2 + F_y^2}$) function of mass coordinates (X, Y). The largest *dot* is the equilibrium position. The smallest *dots* are the Newton-Raphson iterations starting at $x = 0.9\,\text{m}$ and $y = 1.9\,\text{m}$

$$\frac{\partial F_x}{\partial x} = A\frac{l - l_0}{l_0} + A\frac{x^2}{ll_0} \tag{3.9}$$

$$\frac{\partial F_x}{\partial y} = A\frac{xy}{ll_0} \tag{3.10}$$

$$\frac{\partial F_y}{\partial x} = A\frac{yx}{ll_0} \tag{3.11}$$

$$\frac{\partial F_y}{\partial y} = A\frac{l - l_0}{l_0} + A\frac{y^2}{ll_0} \tag{3.12}$$

The Newton-Raphson method accesses the equilibrium solution through iterations. At each iteration the new position is calculated by the following relation:

$$\mathbf{X}_{k+1} = \mathbf{X}_k + \frac{\mathbf{F}(\mathbf{X}_k)}{-F'(\mathbf{X}_k)} \tag{3.13}$$

With:

$$\mathbf{X}_k = \begin{vmatrix} x_k \\ y_k \end{vmatrix} \tag{3.14}$$

$$\mathbf{F}(\mathbf{X}_k) = \begin{vmatrix} F_x(X_k) \\ F_y(X_k) \end{vmatrix} \tag{3.15}$$

The ratio $\frac{\mathbf{F}(\mathbf{X}_k)}{-F'(\mathbf{X}_k)}$ is the displacement $\mathbf{h}$, such as $\mathbf{F}(\mathbf{X}_k) = -F'(\mathbf{X}_k)\mathbf{h}$.

With these equations the equilibrium position is assessed (Fig. 3.5). Figure 3.6 represents the reduction of the force residue with the iterations.

3.1.3 Several Dimensions

3.1.3.1 Main Variables

The positions of the nodes are in vector $\mathbf{X}$, the forces on the nodes are in vector $\mathbf{F}$, and the stiffness matrix is K; x_i and F_i refer to the same node along the same axis.

These variables are as follows:

$$\mathbf{X} = \begin{vmatrix} x_1 \\ x_2 \\ . \\ . \\ x_n \end{vmatrix} \tag{3.16}$$

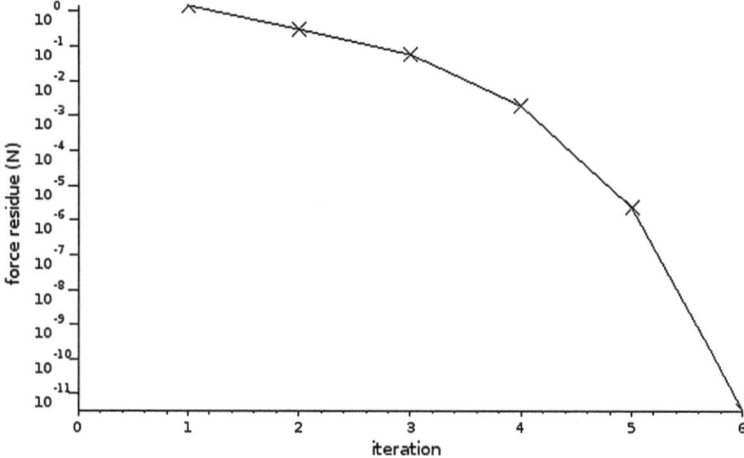

Fig. 3.6 Residue of force ($\sqrt{F_x^2 + F_y^2}$) for each Newton-Raphson method iteration

$$
\mathbf{F} = \begin{vmatrix} F_1 \\ F_2 \\ \cdot \\ \cdot \\ F_n \end{vmatrix}
\tag{3.17}
$$

$$
K = \begin{vmatrix} -\dfrac{\partial F_1}{\partial x_1} & -\dfrac{\partial F_1}{\partial x_2} & \cdots & -\dfrac{\partial F_1}{\partial x_n} \\ -\dfrac{\partial F_2}{\partial x_1} & -\dfrac{\partial F_2}{\partial x_2} & \cdots & -\dfrac{\partial F_2}{\partial x_n} \\ \cdot & \cdot & \cdots & \cdot \\ \cdot & \cdot & & \cdot \\ -\dfrac{\partial F_n}{\partial x_1} & -\dfrac{\partial F_n}{\partial x_2} & \cdots & -\dfrac{\partial F_n}{\partial x_n} \end{vmatrix}
\tag{3.18}
$$

From these three variables the displacement vector (**h**) can be calculated by solving the following system of linear equations:

$$
\mathbf{h}K = \mathbf{F}
\tag{3.19}
$$

3.1.3.2 Iterations

As mentioned earlier, the Newton-Raphson-method is an iterative one. The steps are as follows:

From the position ($\mathbf{X}_k$) of the nodes resulting from iteration k:

$$\mathbf{X}_k = \begin{vmatrix} x_{k1} \\ x_{k2} \\ . \\ . \\ x_{kn} \end{vmatrix} \qquad (3.20)$$

The force $(\mathbf{F}_k)$ on the nodes and the stiffness (K_k) matrix are calculated:

$$\mathbf{F}_k = \begin{vmatrix} F_{k1} \\ F_{k2} \\ . \\ . \\ F_{kn} \end{vmatrix} \qquad (3.21)$$

$$K_k = \begin{vmatrix} K_{k11} & K_{k12} & .. & K_{k1n} \\ K_{k21} & K_{k22} & .. & K_{k2n} \\ . & . & ... & . \\ . & . & ... & . \\ K_{kn1} & K_{kn2} & .. & K_{knn} \end{vmatrix} \qquad (3.22)$$

The node displacements $(\mathbf{h}_k)$ are calculated:

$$\mathbf{h}_k K_k = \mathbf{F}_k \qquad (3.23)$$

The new position of nodes is deduced:

$$\mathbf{X}_{k+1} = \mathbf{X}_k + \mathbf{h}_k \qquad (3.24)$$

3.1.4 Singularity of the Stiffness Matrix

In some cases the stiffness matrix (K) could be singular. In this case solving $\mathbf{h}K = \mathbf{F}$ (Sect. 3.1.3, p. 20) could lead to a very large displacement $(h_i \gg 1)$ and to divergence of the method.

An example can be shown with the unstretched horizontal bar of Fig. 3.7. This bar has two extremities. If the first extremity (on the left on Fig. 3.7) has the horizontal and vertical coordinates $(0, 0)$, the position vector is:

$$\mathbf{X} = \begin{vmatrix} 0 \\ 0 \\ x_3 \\ 0 \end{vmatrix} \qquad (3.25)$$

Fig. 3.7 This bar is articulated around its *left* extremity. A vertical force (F_4) is applied on the *right* extremity. This unstretched bar displays a zero stiffness along the *vertical*

With $x_3 \neq 0$
If the force on the second extremity is vertical, the force vector is:

$$\mathbf{F} = \begin{vmatrix} 0 \\ 0 \\ 0 \\ F_4 \end{vmatrix} \tag{3.26}$$

With $F_4 \neq 0$

As we will see in Sect. 5.2 (p. 71) the stiffness matrix is:

$$K = \begin{vmatrix} K_{11} & 0 & -K_{11} & 0 \\ 0 & 0 & 0 & 0 \\ -K_{11} & 0 & K_{11} & 0 \\ 0 & 0 & 0 & 0 \end{vmatrix} \tag{3.27}$$

The matrix is singular. This is due to the derivative $\frac{\partial F_4}{\partial x_4}$, which is equal to 0 in this case of an unstretched horizontal bar. (i) If the bar is not horizontal this derivative will not be equal to 0, because the derivative of the bar length will not equal 0. (ii) If the bar is in tension (or compression), even horizontal, the derivative $\frac{\partial F_4}{\partial x_4}$ will not equal 0 because the derivative of the tension direction is not equal to 0.

To avoid problems due to singularity, precautions are available, as described below.

3.1.4.1 Additional Stiffness

A simple way is to add an arbitrary value (α) along the diagonal of the stiffness matrix, such that the previous matrix becomes:

$$K = \begin{vmatrix} K_{11}+\alpha & 0 & -K_{11} & 0 \\ 0 & \alpha & 0 & 0 \\ -K_{11} & 0 & K_{11}+\alpha & 0 \\ 0 & 0 & 0 & \alpha \end{vmatrix} \tag{3.28}$$

The added value (α) could decrease along the Newton-Raphson iterations. This added value (α) does not modify the equilibrium position, but only the way to reach this equilibrium.

3.1.4.2 Additional Mechanical Behaviour

Another way to remove singularity is to add further mechanical behaviour. For example, if this bar is in a fluid, air, or water, a vertical displacement will generate a drag in the opposite direction, meaning that the components of the stiffness matrix K_{22} and K_{44} will be not equal to 0.

3.1.4.3 Displacement Limit

A displacement limit could be imposed to avoid too large a value:

$$\mathbf{h}K = \mathbf{F} \tag{3.29}$$

$$if \;\; \mathbf{h}_i > limit \;\; \mathbf{h}_i = limit \tag{3.30}$$
$$if \;\; \mathbf{h}_i \le limit \;\; \mathbf{h}_i = \mathbf{h}_i \tag{3.31}$$

3.2 Other Resolution Methods

3.2.1 Newmark Method

The Newmark method is used to find the equilibrium position of a mechanical structure. The following example in one dimension explains the method in a simplified way.

The method consists first in calculating forces on the structure, then calculating the acceleration on the structure using the dynamic equation ($F = M\gamma$). From this acceleration and using a time step, the speed and the new position of the structure can be calculated [3].

For the example displayed in Fig. 3.1, the equilibrium calculation follows the path shown in Fig. 3.8 with a time step of 0.04 s. Figure 3.9 shows the residue of force. This calculation follows the Newmark explicit method [3].

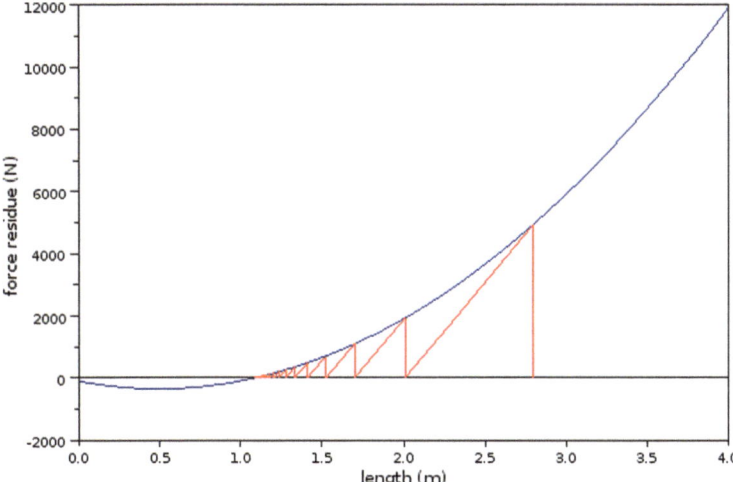

Fig. 3.8 Force on the mass function of spring length and Newmark explicit method iterations

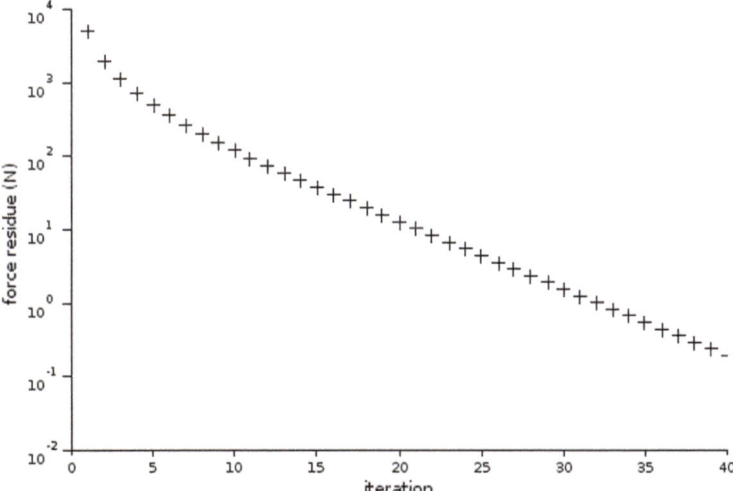

Fig. 3.9 Residue of force for each Newmark explicit method iterations

3.2.2 Energy Minimization

This method consists of finding the position of the structure that leads to the minimum of the energy. The energy involved here is the energy due to the conservative forces only. A conservative force is a force that leads to a variation of energy between two positions independent of the path between these two positions. The main conservative forces involved in marine structures are weight and tension in elastic cables and netting twines.

In these cases the energy between two positions are quite simple to calculate:

$$E_W = W \Delta h \tag{3.32}$$

$$E_T = \frac{1}{2} K \Delta x^2 \tag{3.33}$$

E_W: energy due to the weight (J),
W: weight (N),
Δh: altitude variation between the two positions (m),
E_T: energy due to the tension (J),
K: constant cable stiffness (N/m),
Δx: cable length variation between the two positions (m).

Some forces are not conservative, as in the case of drag force. In such case the energy consumed by the drag depends on the path followed by the structure between the two positions.

Due to non conservative forces, the method of minimization of energy is not quite adapted to solve the equilibrium of marine structures. In case this method is used, the drag forces could be transformed into constant force.

Chapter 4
The Triangular Finite Element for Netting

Abstract The modellings for netting are fully described. The usual modellings based on numerical twines or globalization of twines are partly explained with their limitations. These limitations have drove to the creation of the triangular finite element for netting. This triangular element for netting is fully described. The forces required for the equilibrium calculation are fully described, as well as the stiffness in case of—twines elasticity,—hydrodynamic forces,—twine flexion,—mesh opening stiffness,—fish catch pressure,—inertia,—buoyancy and weight.

Keywords Triangular finite element for netting · Twines tension in netting · Hydrodynamic forces on netting · Twine flexion in netting · Mesh opening stiffness of netting · Fish catch pressure in cod-end

4.1 State-of-the-Art of Numerical Modelling for Nets

4.1.1 Constitutive Law for Nets

There is little or no published work on the constitutive law for nets. Only Rivlin [23], to our knowledge, begins to express the stresses in a net surface, but only under conditions of symmetrical deformation twine. If such constitutive law could be defined, usual finite element softwares could be adapted for nettings.

4.1.2 Twine Numerical Method

The twine numerical method includes almost all the work on numerical modelling of the net [2, 6, 9, 10, 11, 24]. The initial idea is simple: the twines of the net are modelled by bars (called here numerical twines). Then a few adjustments are required.

D. Priour, *A Finite Element Method for Netting*, SpringerBriefs in Environmental Science, DOI: 10.1007/978-94-007-6844-4_4, © The Author(s) 2013

The twines could be modelled by two bars to account for the shortening, which appears as an angle between the bars. The twines could be modelled with a single bar, but Young's modulus in compression is almost zero to account for the shortening. Given the large number of twines in some structures (up to one million), a numerical bar refers to several true twines (Fig. 4.1). This is called globalization.

The major difficulty with this method of globalization lies in the description of the net by numerical twines. Indeed, a structure is very often the assembly of several panels of nets. Therefore, the creation of numerical twines in a panel will generate nodes on its contour. These nodes are the basis for the creation of numerical twines of the adjacent panel (Figs. 4.2 and 4.3).

Figure 4.2a shows four panels (50 by 50 meshes) whose numerical twines connect perfectly (Fig. 4.2b): the nodes on the edges are perfectly aligned with the nodes of the adjacent panels.

Figure 4.3a shows the same example, except that panel 1 is only 45 meshes horizontally. In this case the nodes on the borders do not connect perfectly between panels 4 and 1 (Fig. 4.3b), whereas the connections are perfect on the other three seams. This approach requires facilities such modification of the design of the netting panels. These facilities are not well described in the literature dedicated to this method.

4.2 The Finite Element for Netting

Triangular elements have been developed to model the net (Fig. 4.4). A number of approximations are made in these triangular elements, with the aim of calculating the forces at the vertices of these elements. These are calculated based on the positions

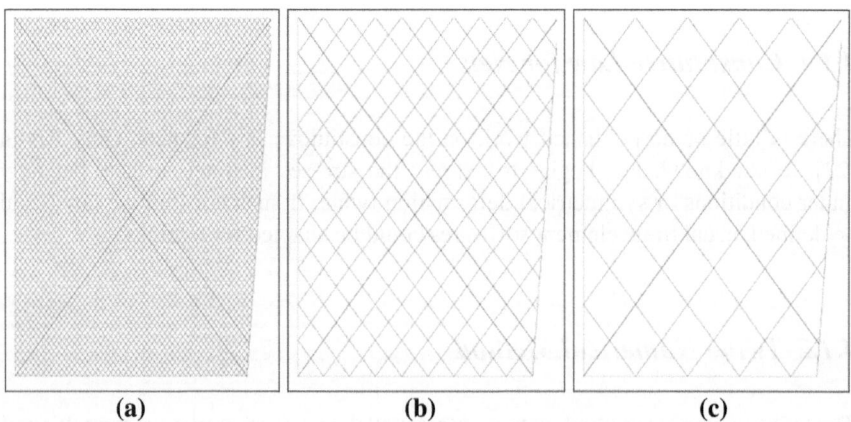

(a)   **(b)**   **(c)**

Fig. 4.1 Control net 50 meshes high by 50 and 45 wide (**a**), with a ratio of globalization of 5 (**b**) and 10 (**c**)

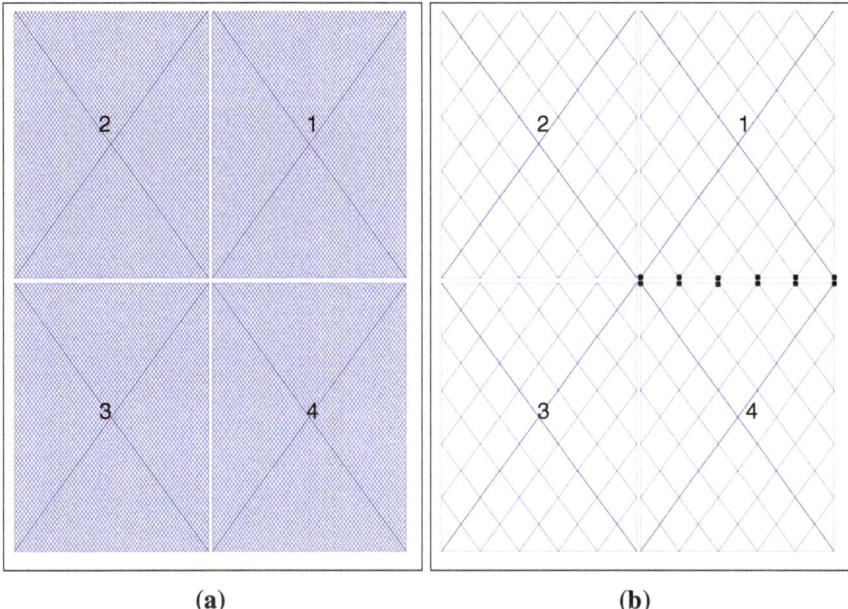

Fig. 4.2 Structure of four panels of 50 by 50 meshes (**a**) discretized in numerical twines (**b**; globalization ratio of 10): the connection between numerical nodes on the borders of panels is perfect (*black dots* for the border between panels)

of the vertices. The basic assumption in modelling nets by triangular elements is that the twines remain parallel. Under these conditions the twines of the same direction have the same deformation. The second assumption is that the twines are modelled as elastic rods.

One difficulty with the method of numerical globalized twines (or numerical twines) was described earlier: nodes on the edges of the panels do not always coincide perfectly (Fig. 4.3b). This difficulty disappears with triangular elements, since the discretization of a netting panel is independent of the discretization of adjacent panels, except on the border. The same panels of Fig. 4.3 are discretized in Fig. 4.5 with triangular elements. Panel 2 in (Fig. 4.5a) is discretized with large triangular elements and in (Fig. 4.5b) with smaller elements. It is clear that triangular element discretization is done very easily, unlike the numerical twines technique. This flexibility in the creation of triangular elements overcomes the cumbersome tool for creating globalized twines. This burden results from many different cases to be processed and consequently adjustments that sometimes make it impossible to fully describe the structure to be studied with the method of numerical twines.

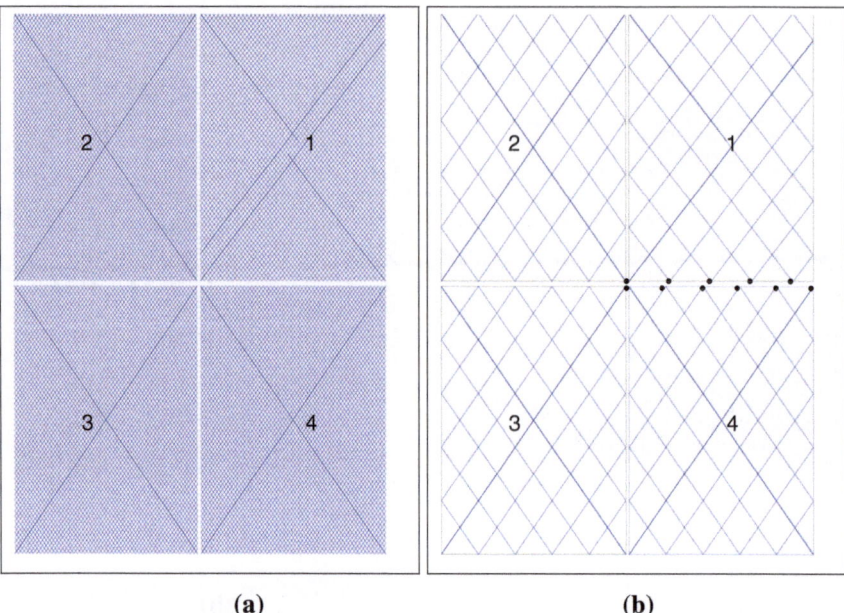

(a)   **(b)**

Fig. 4.3 **a** Four netting panels 50 by 50 meshes except for panel 1, which has only 45 meshes horizontally. **b** The globalization of 10 leads the nodes on the common border of panels 1 and 4 to not connect perfectly: panel 1 has five nodes on its bottom border, while the top border of panel 4 has six nodes (*black dots*)

4.2.1 The Basic Method: Direct Formulation

The triangular finite element dedicated to diamond mesh nets is described here.

The triangular element is defined by its three vertices, which are connected to the net. The coordinates of the vertices in number of twine vectors are then constant, whatever the deformation of the triangle. Figure 4.6 shows an example. In this example the coordinates in twine number of node 1 are 1.5 along the $\mathbf{U}$ twine and -3.5 along the $\mathbf{V}$ twine. It is clear that if the origin of coordinates in twine number changes, the twine coordinates of nodes will change but will not affect the equilibrium position of the net.

These twines are parallel inside the triangular element, which means that the sides of the triangle (**12**, **23**, **31**) are linear combinations of twine vectors ($\mathbf{U}$ and $\mathbf{V}$, cf. Fig. 4.6). This point is the main foundation of the model. These combinations are as follows:

$$\mathbf{12} = (U_2 - U_1)\mathbf{U} + (V_2 - V_1)\mathbf{V} \tag{4.1}$$

$$\mathbf{13} = (U_3 - U_1)\mathbf{U} + (V_3 - V_1)\mathbf{V} \tag{4.2}$$

12 (**13**): vector from vertex 1 (1) to vertex 2 (3).

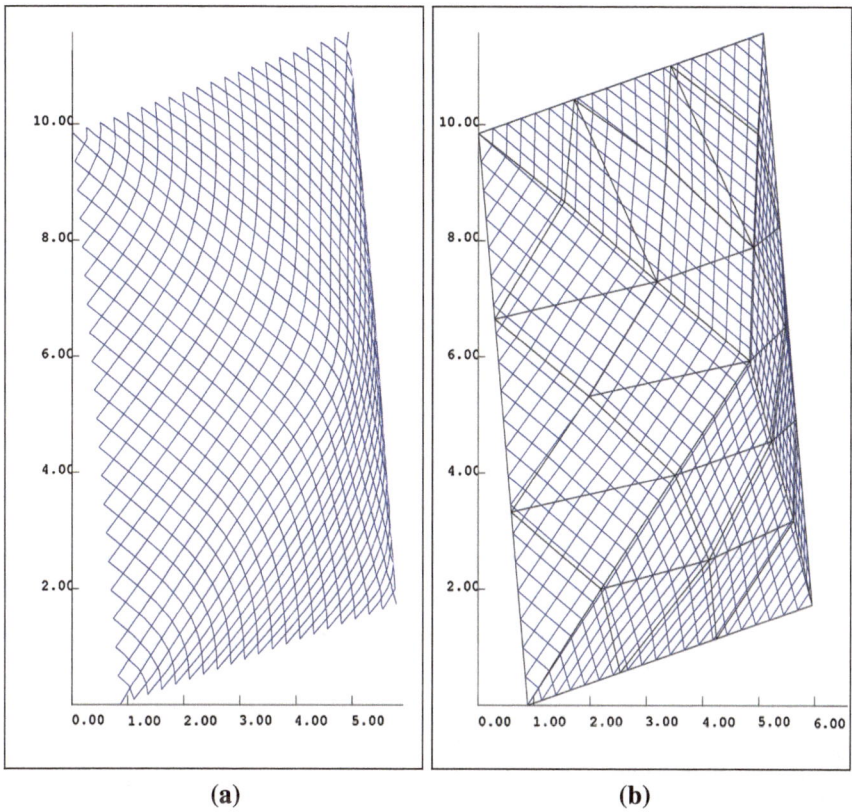

(a)   (b)

Fig. 4.4 The diamond mesh (**a**) is decomposed into triangular elements (**b**). The approximation in each triangle is that twines are parallel and therefore have the same deformation, and that the twines are elastic

The two previous equations with two unknowns (**U** and **V**) then give the following:

$$\mathbf{U} = \frac{V_3 - V_1}{d}\mathbf{12} - \frac{V_2 - V_1}{d}\mathbf{13} \tag{4.3}$$

$$\mathbf{V} = \frac{U_2 - U_1}{d}\mathbf{13} - \frac{U_3 - U_1}{d}\mathbf{12} \tag{4.4}$$

With side vectors:

$$\mathbf{12} = \begin{vmatrix} x_2 - x_1 \\ y_2 - y_1 \\ z_2 - z_1 \end{vmatrix} \tag{4.5}$$

$$\mathbf{13} = \begin{vmatrix} x_3 - x_1 \\ y_3 - y_1 \\ z_3 - z_1 \end{vmatrix} \tag{4.6}$$

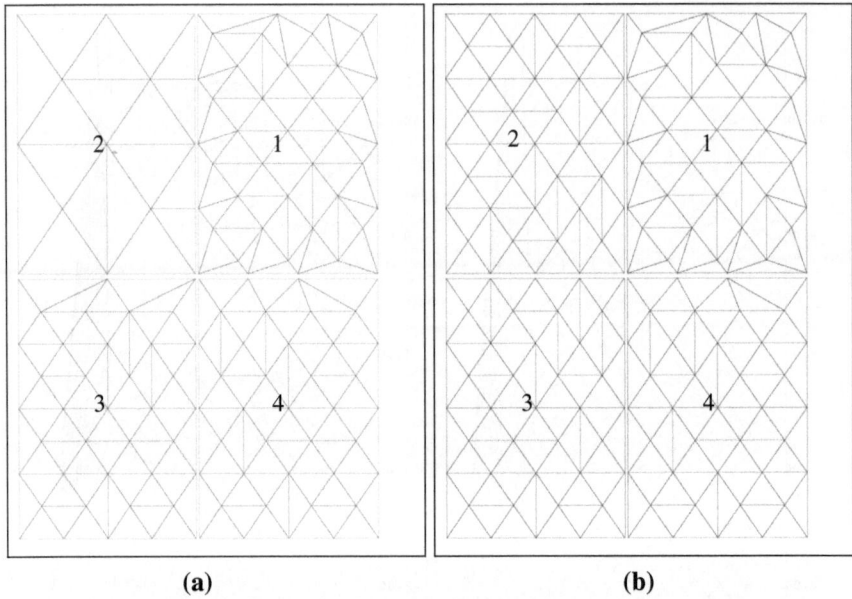

(a)   (b)

Fig. 4.5 Case identical to Fig. 4.3. Although the netting in panel 1 has only 45 meshes horizontally, the triangular element discretization is easy. The step size of panel 2 is larger in (**a**) than in (**b**)

Fig. 4.6 A triangular element: the sides of the triangle
are linear combinations of
twine vectors (**U** and **V**). The
coordinates in twine number
are noted. The origin of theses
coordinates is the intersection
of **U** and **V**

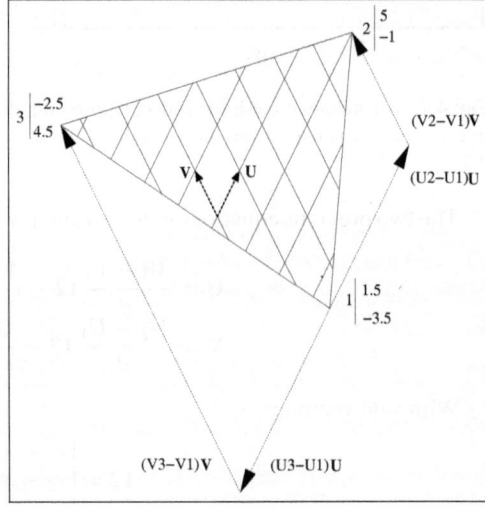

and

$$d = (U_2 - U_1)(V_3 - V_1) - (U_3 - U_1)(V_2 - V_1) \qquad (4.7)$$

x_i, y_i, z_i: Cartesian coordinates of vertex i,
U_i, V_i: coordinates of vertex i in number of twines (twine coordinates).
The twine vectors ($\mathbf{U}$, $\mathbf{V}$) are calculated from the Cartesian coordinates (x_i, y_i, z_i) of the vertices of the triangular element.

It appears that nothing implies that the number of twine coordinates of the vertices of the triangle consists of integers. Therefore, these coordinates can be real. This implies that the vertices of the triangle are not necessarily located on knots of the net (Fig. 4.4). Similarly, nothing prevents the triangle from being smaller than a mesh. It appears that while the triangle does not contain any piece of twine of the net, d is not null, and therefore the triangle contains twines and consequently a deformation energy. In other words, the triangular finite element is a homogenization of the mechanical properties of the net.

It also appears that every point of the twines belongs to only one triangular element and still the same, regardless of the deformation of the net. Points on the contour of a triangular element also belong to the neighbours.

4.2.2 Metric of the Triangular Element

The objective of the finite element method is to calculate the Cartesian coordinates of the numerical nodes. These nodes are, for the netting, the vertices of the triangular elements (Figs. 4.7 and 4.8a).

The nodes are fixed relative to the netting, which means that the coordinates of the nodes in twines or meshes remain constant regardless of the netting deformation.

Figure 4.8b and c show an example of coordinates of a triangular element. Generally speaking, the mesh coordinates are used by the netting maker.

There are relations between the mesh coordinates and the twine coordinates, the bases of which are noted in Fig. 4.8b and c.

The relations between the bases are the following:

$$\mathbf{u} = \mathbf{U} - \mathbf{V} \qquad (4.8)$$
$$\mathbf{v} = \mathbf{U} + \mathbf{V} \qquad (4.9)$$

This leads to:

$$\mathbf{U} = \frac{\mathbf{u} + \mathbf{v}}{2} \qquad (4.10)$$

$$\mathbf{V} = \frac{\mathbf{v} - \mathbf{u}}{2} \qquad (4.11)$$

$\mathbf{u}$, $\mathbf{v}$: mesh coordinates base,
$\mathbf{U}$, $\mathbf{V}$: twine coordinates base.

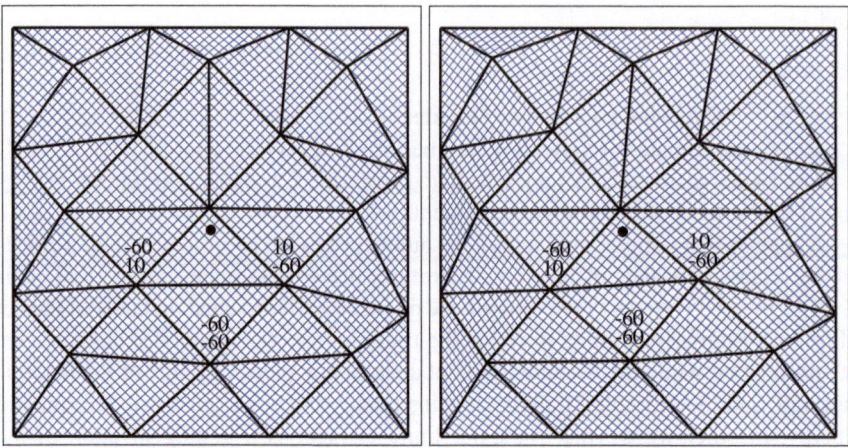

Fig. 4.7 Two deformations of the same structure. The twines coordinates of vertices remain constant. The twines coordinates of three vertices are noted. The *dot* is the origin of twines numbering. Only 1 twine on 5 is drawn

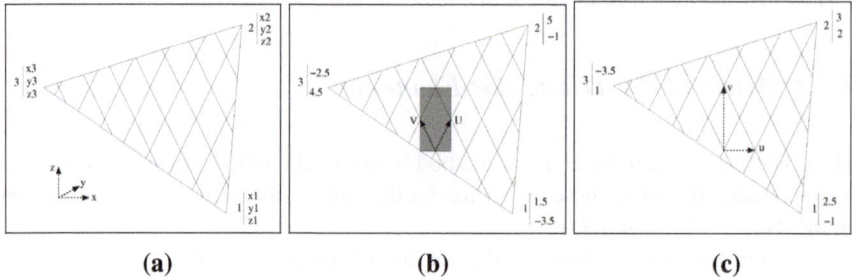

| (a) | (b) | (c) |

Fig. 4.8 Triangular element: Cartesian coordinates (**a**), twines coordinates (**b**), and mesh coordinates (**c**). The *grey* surface is a mesh surface (**b**)

This means that the relations between the twine coordinates and the mesh coordinates of the node P are the following:

$$U_P = u_P + v_P \tag{4.12}$$

$$V_P = v_P - u_P \tag{4.13}$$

and

$$u_P = \frac{U_P - V_P}{2} \tag{4.14}$$

$$v_P = \frac{U_P + V_P}{2} \tag{4.15}$$

Here, U_P and V_P are the twine coordinates, and u_P and v_P are the mesh coordinates of the same node P. In these conditions the vector from origin to node P could be written as follows:

$$\mathbf{OP} = U_P\mathbf{U} + V_P\mathbf{V} \qquad (4.16)$$

$$\mathbf{OP} = u_P\mathbf{u} + v_P\mathbf{v} \qquad (4.17)$$

Because the amplitude of a cross product of vectors is twice the surface of the triangle made of these two vectors, the Cartesian surface of the triangular element (in m^2) is half the amplitude of the cross product of the side vectors of the triangular element:

$$S = \frac{1}{2}|\mathbf{12} \wedge \mathbf{13}| \qquad (4.18)$$

The side vectors in Cartesian coordinates are as follows:

$$\mathbf{12} = \begin{vmatrix} x_2 - x_1 \\ y_2 - y_1 \\ z_2 - z_1 \end{vmatrix} \qquad (4.19)$$

$$\mathbf{13} = \begin{vmatrix} x_3 - x_1 \\ y_3 - y_1 \\ z_3 - z_1 \end{vmatrix} \qquad (4.20)$$

By the same way, the number of meshes, as defined in Fig. 4.8b, is

$$nb_m = \frac{1}{4}|\mathbf{12} \wedge \mathbf{13}| \qquad (4.21)$$

with side vectors in twine coordinates:

$$\mathbf{12} = \begin{vmatrix} U_2 - U_1 \\ V_2 - V_1 \\ 0 \end{vmatrix} \qquad (4.22)$$

$$\mathbf{13} = \begin{vmatrix} U_3 - U_1 \\ V_3 - V_1 \\ 0 \end{vmatrix} \qquad (4.23)$$

The number of meshes in a triangular element is

$$nb_m = \frac{1}{4}[(U_2 - U_1)(V_3 - V_1) - (U_3 - U_1)(V_2 - V_1)] = \frac{d}{4} \qquad (4.24)$$

Because there are two twines U and two twines V per mesh, the number of twines U and V is calculated as follows:

$$nb_U = \frac{d}{2} \tag{4.25}$$

$$nb_V = \frac{d}{2} \tag{4.26}$$

Because there are also two knots per mesh, the number of knots in a triangular element is

$$nb_k = \frac{d}{2} \tag{4.27}$$

The surface (m^2) of one mesh is calculated through the cross product of twines vectors (**U** and **V**):

$$Ms = 2|\mathbf{U} \wedge \mathbf{V}| \tag{4.28}$$

which is also the surface of the triangular element divided by the number of meshes in the element:

$$Ms = \frac{S}{nb_m} \tag{4.29}$$

In the case of Figs. 4.6 and 4.8, $d = 38$, the number of meshes is 9.5, the number of U twines is 18, the number of V twines is 18, and the number of knots is 18.

4.3 The Forces on the Netting

4.3.1 Twine Tension in Diamond Mesh

The tensions in the twines are required to estimate the forces on the vertices due to these tensions. In the hypothesis of linear elasticity, these tensions are deduced from **U** and **V**, which have been previously calculated. In these conditions the twine tensions are as follows:

$$T_u = EA\frac{|\mathbf{U}| - l_0}{l_0} \tag{4.30}$$

$$T_v = EA\frac{|\mathbf{V}| - l_0}{l_0} \tag{4.31}$$

E : Young's modulus of the material (N/m^2),
A : mechanical section of the twines U and V (m^2),
l_o : unstretched length of twine vectors (m).

The principle of virtual work is used here to calculate the forces on the vertices due to the tension in the twines.

The force component along X on vertex 1 of a triangular element is estimated by considering a virtual displacement $(\partial x1)$ along the axis x of vertex 1. This displacement leads to an external work:

$$W_e = F_{x1}\partial x1 \tag{4.32}$$

This displacement also induces a change in the length of mesh bars ($\partial|\mathbf{U}|$ and $\partial|\mathbf{V}|$), an internal work per twine $\partial|\mathbf{U}|T_u$ and $\partial|\mathbf{V}|T_v$ and therefore an internal work for the triangular element:

$$W_i = (\partial|\mathbf{U}|T_u + \partial|\mathbf{V}|T_v)\frac{d}{2} \tag{4.33}$$

The principle of virtual work implies that the external work equals the internal work, since the forces represent the tension in the twines. That gives for each component of force on the three vertices:

$$F_{x1} = \left(T_u\frac{\partial|\mathbf{U}|}{\partial x1} + T_v\frac{\partial|\mathbf{V}|}{\partial x1}\right)\frac{d}{2} \tag{4.34}$$

$$F_{y1} = \left(T_u\frac{\partial|\mathbf{U}|}{\partial y1} + T_v\frac{\partial|\mathbf{V}|}{\partial y1}\right)\frac{d}{2} \tag{4.35}$$

$$F_{z1} = \left(T_u\frac{\partial|\mathbf{U}|}{\partial z1} + T_v\frac{\partial|\mathbf{V}|}{\partial z1}\right)\frac{d}{2} \tag{4.36}$$

$$F_{x2} = \left(T_u\frac{\partial|\mathbf{U}|}{\partial x2} + T_v\frac{\partial|\mathbf{V}|}{\partial x2}\right)\frac{d}{2} \tag{4.37}$$

$$F_{y2} = \left(T_u\frac{\partial|\mathbf{U}|}{\partial y2} + T_v\frac{\partial|\mathbf{V}|}{\partial y2}\right)\frac{d}{2} \tag{4.38}$$

$$F_{z2} = \left(T_u\frac{\partial|\mathbf{U}|}{\partial z2} + T_v\frac{\partial|\mathbf{V}|}{\partial z2}\right)\frac{d}{2} \tag{4.39}$$

$$F_{x3} = \left(T_u\frac{\partial|\mathbf{U}|}{\partial x3} + T_v\frac{\partial|\mathbf{V}|}{\partial x3}\right)\frac{d}{2} \tag{4.40}$$

$$F_{y3} = \left(T_u\frac{\partial|\mathbf{U}|}{\partial y3} + T_v\frac{\partial|\mathbf{V}|}{\partial y3}\right)\frac{d}{2} \tag{4.41}$$

$$F_{z3} = \left(T_u\frac{\partial|\mathbf{U}|}{\partial z3} + T_v\frac{\partial|\mathbf{V}|}{\partial z3}\right)\frac{d}{2} \tag{4.42}$$

The derivatives $\frac{\partial|\mathbf{U}|}{\partial x1} \dots \frac{\partial|\mathbf{V}|}{\partial z3}$ can be calculated, as the equations relating to U, V and X_i, Y_i, Z_i have already been described. This gives the following vectors force for the three vertices:

$$\mathbf{F_1} = (V_3 - V_2)T_u\frac{\mathbf{U}}{2|\mathbf{U}|} + (U_2 - U_3)T_v\frac{\mathbf{V}}{2|\mathbf{V}|} \tag{4.43}$$

$$\mathbf{F_2} = (V_1 - V_3)T_u\frac{\mathbf{U}}{2|\mathbf{U}|} + (U_3 - U_1)T_v\frac{\mathbf{V}}{2|\mathbf{V}|} \tag{4.44}$$

$$\mathbf{F_3} = (V_2 - V_1)T_u\frac{\mathbf{U}}{2|\mathbf{U}|} + (U_1 - U_2)T_v\frac{\mathbf{V}}{2|\mathbf{V}|} \tag{4.45}$$

The Newton-Raphson method, described earlier, requires the calculation of the stiffness matrix, which is calculated from the derivatives of effort with respect to the positions of the vertices of the triangular element. The 81 derivatives, that is to say, by 9 by 9 component coordinates, are then the following:

The stiffness matrix:

$$K = \begin{pmatrix} -\frac{\partial F_{x1}}{\partial x_1} & -\frac{\partial F_{x1}}{\partial y_1} & \cdots & -\frac{\partial F_{x1}}{\partial z_3} \\ -\frac{\partial F_{y1}}{\partial x_1} & -\frac{\partial F_{y1}}{\partial y_1} & \cdots & -\frac{\partial F_{y1}}{\partial z_3} \\ \cdot & \cdot & \cdots & \cdot \\ \cdot & \cdot & \cdots & \cdot \\ \cdot & \cdot & \cdots & \cdot \\ -\frac{\partial F_{z3}}{\partial x_1} & -\frac{\partial F_{z3}}{\partial y_1} & \cdots & -\frac{\partial F_{z3}}{\partial z_3} \end{pmatrix} \tag{4.46}$$

The components are calculated as follows:

$$\frac{\partial F_{w1}}{\partial t} = \frac{EA_u(V_3 - V_2)}{2}\left[\frac{\partial U_w}{\partial t}\left(\frac{1}{n_0} - \frac{1}{|\mathbf{U}|}\right) + \frac{\partial|\mathbf{U}|}{\partial t}\frac{U_w}{|\mathbf{U}|^2}\right]$$
$$+ \frac{EA_v(U_2 - U_3)}{2}\left[\frac{\partial V_w}{\partial t}\left(\frac{1}{n_0} - \frac{1}{|\mathbf{V}|}\right) + \frac{\partial|\mathbf{V}|}{\partial t}\frac{V_w}{|\mathbf{V}|^2}\right] \tag{4.47}$$

$$\frac{\partial F_{w2}}{\partial t} = \frac{EA_u(V_1 - V_3)}{2}\left[\frac{\partial U_w}{\partial t}\left(\frac{1}{n_0} - \frac{1}{|\mathbf{U}|}\right) + \frac{\partial|\mathbf{U}|}{\partial t}\frac{U_w}{|\mathbf{U}|^2}\right]$$
$$+ \frac{EA_v(U_3 - U_1)}{2}\left[\frac{\partial V_w}{\partial t}\left(\frac{1}{n_0} - \frac{1}{|\mathbf{V}|}\right) + \frac{\partial|\mathbf{V}|}{\partial t}\frac{V_w}{|\mathbf{V}|^2}\right] \tag{4.48}$$

$$\frac{\partial F_{w3}}{\partial t} = \frac{EA_u(V_2 - V_1)}{2}\left[\frac{\partial U_w}{\partial t}\left(\frac{1}{n_0} - \frac{1}{|\mathbf{U}|}\right) + \frac{\partial|\mathbf{U}|}{\partial t}\frac{U_w}{|\mathbf{U}|^2}\right]$$
$$+ \frac{EA_v(U_1 - U_2)}{2}\left[\frac{\partial V_w}{\partial t}\left(\frac{1}{n_0} - \frac{1}{|\mathbf{V}|}\right) + \frac{\partial|\mathbf{V}|}{\partial t}\frac{V_w}{|\mathbf{V}|^2}\right] \tag{4.49}$$

With:

$w = x, y, z,$

$t = x1, y1, z1, x2, y2, z2, x3, y3, z3.$

The following derivatives are also required.

The derivatives of the components of **U** are as follows:

$$\frac{\partial U_x}{\partial x1} = \frac{\partial U_y}{\partial y1} = \frac{\partial U_z}{\partial z1} = \frac{V_2 - V_3}{d} \tag{4.50}$$

$$\frac{\partial U_x}{\partial x2} = \frac{\partial U_y}{\partial y2} = \frac{\partial U_z}{\partial z2} = \frac{V_3 - V_1}{d} \tag{4.51}$$

$$\frac{\partial U_x}{\partial x3} = \frac{\partial U_y}{\partial y3} = \frac{\partial U_z}{\partial z3} = \frac{V_1 - V_2}{d} \tag{4.52}$$

$$\frac{\partial U_x}{\partial yi} = \frac{\partial U_x}{\partial zi} = \frac{\partial U_y}{\partial zi} = \frac{\partial U_y}{\partial xi} = \frac{\partial U_z}{\partial xi} = \frac{\partial U_z}{\partial yi} = 0 \tag{4.53}$$

The derivatives of the components of **V** are the following:

$$\frac{\partial V_x}{\partial x1} = \frac{\partial V_y}{\partial y1} = \frac{\partial V_z}{\partial z1} = \frac{U_3 - U_2}{d} \tag{4.54}$$

$$\frac{\partial V_x}{\partial x2} = \frac{\partial V_y}{\partial y2} = \frac{\partial V_z}{\partial z2} = \frac{U_1 - U_3}{d} \tag{4.55}$$

$$\frac{\partial V_x}{\partial x3} = \frac{\partial V_y}{\partial y3} = \frac{\partial V_z}{\partial z3} = \frac{U_2 - U_1}{d} \tag{4.56}$$

$$\frac{\partial V_x}{\partial yi} = \frac{\partial V_x}{\partial zi} = \frac{\partial V_y}{\partial zi} = \frac{\partial V_y}{\partial xi} = \frac{\partial V_z}{\partial xi} = \frac{\partial V_z}{\partial yi} = 0 \tag{4.57}$$

The derivatives of |**U**| follow:

$$\frac{\partial |U|}{\partial x1} = \frac{V_2 - V_3}{d^2} [(x_2 - x_1)(V_3 - V_1) - (x_3 - x_1)(V_2 - V_1)] \tag{4.58}$$

$$\frac{\partial |U|}{\partial x2} = \frac{V_3 - V_1}{d^2} [(x_2 - x_1)(V_3 - V_1) - (x_3 - x_1)(V_2 - V_1)] \tag{4.59}$$

$$\frac{\partial |U|}{\partial x3} = \frac{V_1 - V_2}{d^2} [(x_2 - x_1)(V_3 - V_1) - (x_3 - x_1)(V_2 - V_1)] \tag{4.60}$$

$$\frac{\partial |U|}{\partial y1} = \frac{V_2 - V_3}{d^2} [(y_2 - y_1)(V_3 - V_1) - (y_3 - y_1)(V_2 - V_1)] \tag{4.61}$$

$$\frac{\partial |U|}{\partial y2} = \frac{V_3 - V_1}{d^2} [(y_2 - y_1)(V_3 - V_1) - (y_3 - y_1)(V_2 - V_1)] \tag{4.62}$$

$$\frac{\partial |U|}{\partial y3} = \frac{V_1 - V_2}{d^2} [(y_2 - y_1)(V_3 - V_1) - (y_3 - y_1)(V_2 - V_1)] \tag{4.63}$$

$$\frac{\partial |U|}{\partial z1} = \frac{V_2 - V_3}{d^2} [(z_2 - z_1)(V_3 - V_1) - (z_3 - z_1)(V_2 - V_1)] \tag{4.64}$$

$$\frac{\partial |U|}{\partial z2} = \frac{V_3 - V_1}{d^2} [(z_2 - z_1)(V_3 - V_1) - (z_3 - z_1)(V_2 - V_1)] \tag{4.65}$$

$$\frac{\partial |U|}{\partial z3} = \frac{V_1 - V_2}{d^2} [(z_2 - z_1)(V_3 - V_1) - (z_3 - z_1)(V_2 - V_1)] \tag{4.66}$$

The derivatives of $|\mathbf{V}|$ are shown below:

$$\frac{\partial |\mathbf{V}|}{\partial x1} = \frac{U_2 - U_3}{d^2} \left[(x_2 - x_1)(U_3 - U_1) - (x_3 - x_1)(U_2 - U_1) \right] \quad (4.67)$$

$$\frac{\partial |\mathbf{V}|}{\partial x2} = \frac{U_3 - U_1}{d^2} \left[(x_2 - x_1)(U_3 - U_1) - (x_3 - x_1)(U_2 - U_1) \right] \quad (4.68)$$

$$\frac{\partial |\mathbf{V}|}{\partial x3} = \frac{U_1 - U_2}{d^2} \left[(x_2 - x_1)(U_3 - U_1) - (x_3 - x_1)(U_2 - U_1) \right] \quad (4.69)$$

$$\frac{\partial |\mathbf{V}|}{\partial y1} = \frac{U_2 - U_3}{d^2} \left[(y_2 - y_1)(U_3 - U_1) - (y_3 - y_1)(U_2 - U_1) \right] \quad (4.70)$$

$$\frac{\partial |\mathbf{V}|}{\partial y2} = \frac{U_3 - U_1}{d^2} \left[(y_2 - y_1)(U_3 - U_1) - (y_3 - y_1)(U_2 - U_1) \right] \quad (4.71)$$

$$\frac{\partial |\mathbf{V}|}{\partial y3} = \frac{U_1 - U_2}{d^2} \left[(y_2 - y_1)(U_3 - U_1) - (y_3 - y_1)(U_2 - U_1) \right] \quad (4.72)$$

$$\frac{\partial |\mathbf{V}|}{\partial z1} = \frac{U_2 - U_3}{d^2} \left[(z_2 - z_1)(U_3 - U_1) - (z_3 - z_1)(U_2 - U_1) \right] \quad (4.73)$$

$$\frac{\partial |\mathbf{V}|}{\partial z2} = \frac{U_3 - U_1}{d^2} \left[(z_2 - z_1)(U_3 - U_1) - (z_3 - z_1)(U_2 - U_1) \right] \quad (4.74)$$

$$\frac{\partial |\mathbf{V}|}{\partial z3} = \frac{U_1 - U_2}{d^2} \left[(z_2 - z_1)(U_3 - U_1) - (z_3 - z_1)(U_2 - U_1) \right] \quad (4.75)$$

4.3.2 Twine Tension in Hexagonal Mesh

The same technique for the diamond mesh netting is used for hexagonal ones. The triangular element dedicated to the hexagonal mesh netting has the same assumption as previously adopted: the three families of twines inside the element are parallel, i.e., $\mathbf{l}$, $\mathbf{m}$, and $\mathbf{n}$ twine vectors, are parallel (Fig. 4.9).

The mesh base (shaded area in Fig. 4.9) is first defined. This base mesh is defined as a parallelogram; its corners coincide with knots, and it includes two $\mathbf{l}$ twine vectors, two $\mathbf{m}$ twine vectors, and two $\mathbf{n}$ twine vectors. This base mesh is also used to quantify the number of meshes inside the triangular element. The vertices of the triangular element then have coordinates in base meshes (U_1, U_2, U_3, V_1, V_2, V_3; Fig. 4.9).

Vectors $\mathbf{U}$ and $\mathbf{V}$ are the sides of the mesh base. There are linear relations between these two vectors and the sides of the triangular element (arrows on Fig. 4.9):

$$\mathbf{12} = (U_2 - U_1)\mathbf{U} + (V_2 - V_1)\mathbf{V} \quad (4.76)$$

$$\mathbf{13} = (U_3 - U_1)\mathbf{U} + (V_3 - V_1)\mathbf{V} \quad (4.77)$$

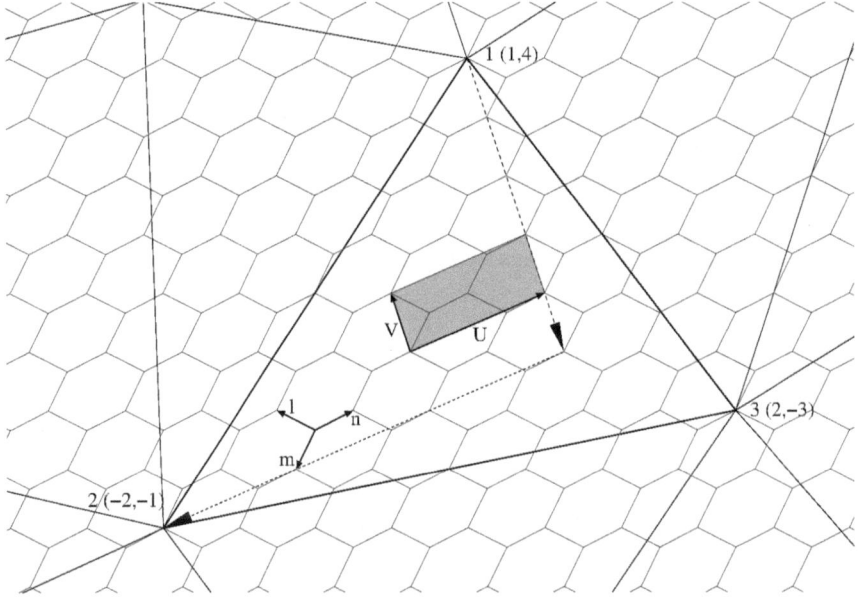

Fig. 4.9 Triangular element dedicated to the hexagonal mesh nets. The twine vectors are **l**, **m**, and **n**. The number of meshes are noted for each vertex. The mesh base is in *grey* and is defined by vectors **U** and **V**

The two previous equations give the following as in the case of diamond mesh (see Sect. 4.2.1, page 30), namely:

$$\mathbf{U} = \frac{V_3 - V_1}{d}\mathbf{12} - \frac{V_2 - V_1}{d}\mathbf{13} \tag{4.78}$$

$$\mathbf{V} = \frac{U_3 - U_1}{d}\mathbf{12} - \frac{U_2 - U_1}{d}\mathbf{13} \tag{4.79}$$

With vectors of the sides of the mesh base:

$$\mathbf{12} = \begin{vmatrix} x_2 - x_1 \\ y_2 - y_1 \\ z_2 - z_1 \end{vmatrix} \tag{4.80}$$

$$\mathbf{13} = \begin{vmatrix} x_3 - x_1 \\ y_3 - y_1 \\ z_3 - z_1 \end{vmatrix} \tag{4.81}$$

and

$$d = (U_2 - U_1)(V_3 - V_1) - (U_3 - U_1)(V_2 - V_1) \tag{4.82}$$

x_i, y_i, z_i: Cartesian coordinates of vertex i.

The number of base meshes in a triangular element is equal to $d/2$, the total number twine vectors is $3d$, the number of twine vectors $\mathbf{l}$, $\mathbf{m}$, or $\mathbf{n}$ is d, and the number of nodes is $2d$.

Tensions in twine vectors $\mathbf{l}$, $\mathbf{m}$, and $\mathbf{n}$ are now calculated. This is done by solving the force balance of the twines. This is solved by writing the following equations:

(1) The base mesh definition leads to (Fig. 4.9) :

$$\mathbf{U} = -\mathbf{m} + 2\mathbf{n} - \mathbf{l} \tag{4.83}$$
$$\mathbf{V} = -\mathbf{m} + \mathbf{l} \tag{4.84}$$

(2) The amplitude of tension in the twines gives:

$$|\mathbf{T}_l| = E A_l \frac{|\mathbf{l}| - l_0}{l_0} \tag{4.85}$$

$$|\mathbf{T}_m| = E A_m \frac{|\mathbf{m}| - m_0}{m_0} \tag{4.86}$$

$$|\mathbf{T}_n| = E A_n \frac{|\mathbf{n}| - n_0}{n_0} \tag{4.87}$$

(3) The balance of tensions leads to:

$$\mathbf{T}_l + \mathbf{T}_m + \mathbf{T}_n = 0 \tag{4.88}$$

This gives six equations with six unknowns ($\mathbf{l}$, $\mathbf{m}$, $\mathbf{n}$, $\mathbf{T}_l$, $\mathbf{T}_m$, $\mathbf{T}_n$).

4.3.2.1 Equilibrium of the Joint Knot

The six previous equations can be reduced to the two that follow with two unknowns (m_x and m_y components of $\mathbf{m}$), since the triangular element has been turned in the plane XOY [17, 19]:

$$\frac{m_x + V_x}{\sqrt{(m_x + V_x)^2 + (m_y + V_y)^2}} \frac{E_l A_l}{l_o} \left[\sqrt{(m_x + V_x)^2 + (m_y + V_y)^2} - l_o \right]$$

$$+ \frac{m_x}{\sqrt{m_x^2 + m_y^2}} \frac{E_m A_m}{m_o} \left[\sqrt{m_x^2 + m_y^2} - m_o \right]$$

$$+ \frac{m_x + \frac{U_x + V_x}{2}}{\sqrt{\left(m_x + \frac{U_x + V_x}{2}\right)^2 + \left(m_y + \frac{U_y + V_y}{2}\right)^2}} \frac{E_n A_n}{n_o}$$

$$\times \left[\sqrt{\left(m_x + \frac{U_x + V_x}{2}\right)^2 + \left(m_y + \frac{U_y + V_y}{2}\right)^2} - n_o \right]$$
$$= 0 \tag{4.89}$$

$$\frac{m_y + V_y}{\sqrt{(m_x + V_x)^2 + (m_y + V_y)^2}} \frac{E_l A_l}{l_o} \left[\sqrt{(m_y + V_y)^2 + (m_y + V_y)^2} - l_o \right]$$

$$+ \frac{m_y}{\sqrt{m_x^2 + m_y^2}} \frac{E_m A_m}{m_o} \left[\sqrt{m_y^2 + m_y^2} - m_o \right]$$

$$+ \frac{m_y + \frac{U_y + V_y}{2}}{\sqrt{\left(m_x + \frac{U_x + V_x}{2}\right)^2 + \left(m_y + \frac{U_y + V_y}{2}\right)^2}} \frac{E_n A_n}{n_o}$$

$$\times \left[\sqrt{\left(m_y + \frac{U_y + V_y}{2}\right)^2 + \left(m_y + \frac{U_y + V_y}{2}\right)^2} - n_o \right]$$
$$= 0 \tag{4.90}$$

m_x, m_y: components of m twine (m),
l_o, m_o, n_o: unstretched length of twines l, m, and n (m),
U_x, U_y, V_x, V_y: components of the sides of the mesh base (m; see Fig. 4.9),
E_l, E_m, E_n: Young modulus of twines l, m, and n (Pa),
A_l, A_m, A_n: section of twines l, m, and n (m^2).
These two equations describe the equilibrium of the joint knot of three twines in a triangle, the sides of which are $\frac{U+V}{2}$ and V (Fig. 4.10). These equations are in newtons.

4.3.2.2 Approximation of the Equilibrium of the Joint

The analytical solution of the two previous equations has not been found. Therefore, the following approximation has been made to simplify the equations. This approximation is acceptable because the stretched lengths of the twines are close to the unstretched length.

$$\frac{m_x}{|\mathbf{m}|} \approx \frac{m_x}{m_o} \tag{4.91}$$

$$\frac{m_y}{|\mathbf{m}|} \approx \frac{m_y}{m_o} \tag{4.92}$$

Fig. 4.10 The three twines
are in the triangle defined by
$\frac{U+V}{2}$ and **V** (cf. Fig. 4.9)

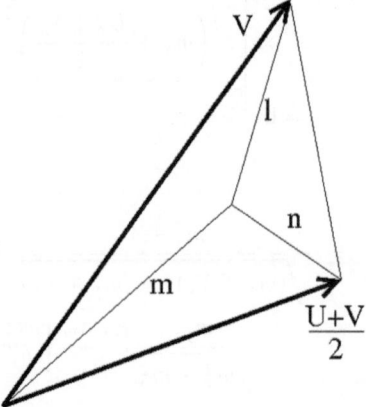

With this approximation the two previous equilibrium equations are reduced to
the following:

$$(m_x + V_x)\frac{E_l A_l}{l_o^2}\left(\sqrt{(m_x + V_x)^2 + (m_y + V_y)^2} - l_o\right) + m_x \frac{E_m A_m}{m_o^2}\left(\sqrt{m_x^2 + m_y^2} - m_o\right)$$

$$+ \left(m_x + \frac{U_x + V_x}{2}\right)\frac{E_n A_n}{n_o^2}\left(\sqrt{\left(m_x + \frac{U_x + V_x}{2}\right)^2 + \left(m_y + \frac{U_y + V_y}{2}\right)^2} - n_o\right) = 0$$

$$\tag{4.93}$$

$$(m_y + V_y)\frac{E_l A_l}{l_o^2}\left(\sqrt{(m_x + V_x)^2 + (m_y + V_y)^2} - l_o\right) + m_y \frac{E_m A_m}{m_o^2}\left(\sqrt{m_x^2 + m_y^2} - m_o\right)$$

$$+ \left(m_y + \frac{U_y + V_y}{2}\right)\frac{E_n A_n}{n_o^2}\left(\sqrt{\left(m_x + \frac{U_x + V_x}{2}\right)^2 + \left(m_y + \frac{U_y + V_y}{2}\right)^2} - n_o\right) = 0$$

$$\tag{4.94}$$

They are the complete form of the following:

$$l_x \frac{E_l A_l}{l_o^2}(|\mathbf{l}| - l_o) + m_x \frac{E_m A_m}{m_o^2}(|\mathbf{m}| - m_o) + n_x \frac{E_n A_n}{n_o^2}(|\mathbf{n}| - n_o) = 0 \tag{4.95}$$

$$l_y \frac{E_l A_l}{l_o^2}(|\mathbf{l}| - l_o) + m_y \frac{E_m A_m}{m_o^2}(|\mathbf{m}| - m_o) + n_y \frac{E_n A_n}{n_o^2}(|\mathbf{n}| - n_o) = 0 \tag{4.96}$$

4.3.2.3   Newton-Raphson Method

The previous approximation has not been sufficient to reach the analytical solution.
The Newton-Raphson method is used to find a numerical solution [4].

For each iteration the displacement h is searched to find the equilibrium:

$$h_k = \frac{F(x_k)}{-F'(x_k)} \tag{4.97}$$

$$x_{k+1} = x_k + h_k \tag{4.98}$$

k: iteration number,
F: force on nodes,
x: position of nodes.
Here:

$$F = \begin{cases} l_x \frac{E_l A_l}{l_o^2}(|\mathbf{l}| - l_o) + m_x \frac{E_m A_m}{m_o^2}(|\mathbf{m}| - m_o) + n_x \frac{E_n A_n}{n_o^2}(|\mathbf{n}| - n_o) = F_1 \\ l_y \frac{E_l A_l}{l_o^2}(|\mathbf{l}| - l_o) + m_y \frac{E_m A_m}{m_o^2}(|\mathbf{m}| - m_o) + n_y \frac{E_n A_n}{n_o^2}(|\mathbf{n}| - n_o) = F_2 \end{cases} \tag{4.99}$$

$$\mathbf{x} = \begin{cases} m_x \\ m_y \end{cases} \tag{4.100}$$

The derivative is:

$$F' = \begin{vmatrix} D_{11} & D_{12} \\ D_{21} & D_{22} \end{vmatrix}. \tag{4.101}$$

With:

$$D_{11} = -\left[\frac{EA_l}{l_o^2}\left(1 - l_o + \frac{l_x^2}{\mathbf{l}}\right) + \frac{EA_m}{m_o^2}\left(\mathbf{m} - m_o + \frac{m_x^2}{\mathbf{m}}\right) + \frac{EA_n}{n_o^2}\left(\mathbf{n} - n_o + \frac{n_x^2}{\mathbf{n}}\right)\right] \tag{4.102}$$

$$D_{12} = D_{21} = -\left[\frac{EA_l}{l_o^2}\frac{l_x l_y}{\mathbf{l}} + \frac{EA_m}{m_o^2}\frac{m_x m_y}{\mathbf{m}} + \frac{EA_n}{n_o^2}\frac{n_x n_y}{\mathbf{n}}\right] \tag{4.103}$$

$$D_{22} = -\left[\frac{EA_l}{l_o^2}\left(1 - l_o + \frac{l_y^2}{\mathbf{l}}\right) + \frac{EA_m}{m_o^2}\left(\mathbf{m} - m_o + \frac{m_y^2}{\mathbf{m}}\right) + \frac{EA_n}{n_o^2}\left(\mathbf{n} - n_o + \frac{n_y^2}{\mathbf{n}}\right)\right] \tag{4.104}$$

With the previous conditions the displacement (**h**) can be calculated:

$$\mathbf{h} = \begin{cases} \frac{D_{22}F_1 - D_{12}F_2}{D_{22}D_{11} - D_{12}D_{21}} \\ \frac{D_{22}F_2 - D_{21}F_1}{D_{22}D_{11} - D_{12}D_{21}} \end{cases} \tag{4.105}$$

4.3.2.4 Forces on Nodes

The forces on the sides of the triangular element are calculated from the twine tension. These forces are related to the number of twines through the sides of the triangle.

This number of twines through each side can be calculated based on the number of base mesh of each vertex.

The effort on the side along $\mathbf{U}$ of the base mesh (Fig. 4.9) is

$$\mathbf{F}_U = \mathbf{T}_l - \mathbf{T}_m \tag{4.106}$$

The effort along $\mathbf{V}$ is

$$\mathbf{F}_V = -\mathbf{T}_n \tag{4.107}$$

Under these conditions, the effort on each side of the triangle can be deduced:

$$\mathbf{T}_{12} = (U_2 - U_1)(\mathbf{T}_l - \mathbf{T}_m) + (V_2 - V_1)(-\mathbf{T}_n) \tag{4.108}$$
$$\mathbf{T}_{23} = (U_3 - U_2)(\mathbf{T}_l - \mathbf{T}_m) + (V_3 - V_2)(-\mathbf{T}_n) \tag{4.109}$$
$$\mathbf{T}_{31} = (U_1 - U_3)(\mathbf{T}_l - \mathbf{T}_m) + (V_1 - V_3)(-\mathbf{T}_n) \tag{4.110}$$

Here, $\mathbf{T}_{ij}$ is the effort on the side ij of the triangular element.

Each side effort is distributed on each end of this side as the twines are evenly distributed along the sides of the triangle:

$$\mathbf{F}_1 = \frac{\mathbf{T}_{12} + \mathbf{T}_{31}}{2} \tag{4.111}$$

$$\mathbf{F}_2 = \frac{\mathbf{T}_{23} + \mathbf{T}_{12}}{2} \tag{4.112}$$

$$\mathbf{F}_3 = \frac{\mathbf{T}_{31} + \mathbf{T}_{23}}{2} \tag{4.113}$$

$\mathbf{F}_1$, $\mathbf{F}_2$, and $\mathbf{F}_3$ are the forces on the three vertices of the triangular element due to the tension in the twines.

The contribution of the stiffness matrix is not described here.

4.3.3 Hydrodynamic Drag

4.3.3.1 Introduction

The drag force on the netting is calculated in this model as the sum of the drag force on each twine ($\mathbf{U}$ and $\mathbf{V}$). This assumption is probably questionable, because the drag on a twine alone is surely not exactly the same as the drag on this twine among other twines as it is the case in a netting. Anyway, this assumption leads to the calculation of the drag of each triangular element because for each the twines vectors are known, as described earlier. The formulation for the twine vector drag is based on the assumptions of Morrison adapted by Landweber and Richtmeyer [8, 22].

Fig. 4.11 Normal (**F**) and
tangential (**T**) forces on a
twine due to the relative
velocity of water (**c**)

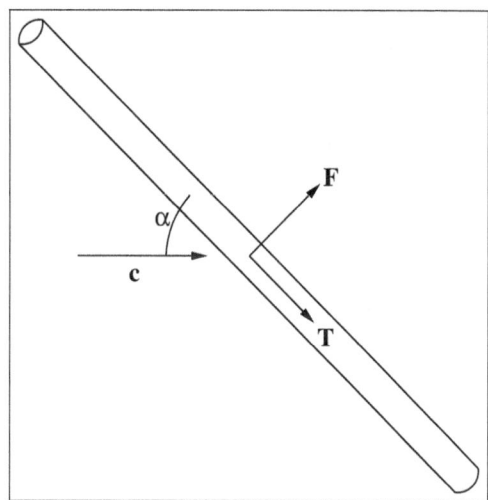

The drag amplitudes on the U twines used in the model (Fig. 4.11) are:

$$|\mathbf{F}| = \frac{1}{2}\rho C_d D l_0 \left[|\mathbf{c}|sin(\alpha)\right]^2 \frac{d}{2} \tag{4.114}$$

$$|\mathbf{T}| = f\frac{1}{2}\rho C_d D l_0 \left[|\mathbf{c}|cos(\alpha)\right]^2 \frac{d}{2} \tag{4.115}$$

The directions of the drag on the **U** twine vectors are:

$$\frac{\mathbf{F}}{|\mathbf{F}|} = \frac{\mathbf{U} \wedge (\mathbf{c} \wedge \mathbf{U})}{|\mathbf{U} \wedge (\mathbf{c} \wedge \mathbf{U})|} \tag{4.116}$$

$$\frac{\mathbf{T}}{|\mathbf{T}|} = \frac{\mathbf{F} \wedge (\mathbf{c} \wedge \mathbf{F})}{|\mathbf{F} \wedge (\mathbf{c} \wedge \mathbf{F})|} \tag{4.117}$$

F: normal drag (N) on the U twines, following the assumptions of Landweber,
T: tangential drag (N) on the U twines, Richtmeyer hypothesis,
ρ: density of water (kg/m³),
C_d: normal drag coefficient,
f: tangential drag coefficient,
D: diameter of twine (m),
l_0: length of twine vector (m),
c: water velocity relative to the twine (m/s),
α: angle between the U twine and the water velocity (radians),
$d/2$: number of U twine vectors in the triangular element.

In the equations of drag amplitude, the expressions $|\mathbf{c}|sin(\alpha)$ and $|\mathbf{c}|cos(\alpha)$ are
the normal and tangential projections on **c** along the U twine vector.

The drag on V twines for a triangular element are similar: **U** is replaced by **V** and
α by β.

The length of twine vectors used in the formulation of drag amplitude can be assessed by $|U|$ for the U twines and by $|V|$ for the V twines. That would mean it takes into account the twine elongation. Generally speaking, a twine elongation is associated with a diameter D reduction by the Poisson coefficient. Because this Poisson coefficient is not taken into account in the present modelling, the twine surface is approximated by Dl_0, where D is the diameter of the twines and l_0 is the unstretched length of the twine vectors.

All parameters, including the angles α and β, are constant and known for each triangular element. Therefore, the drag can be calculated for each triangular element. The drag force for a triangular element is spread over the three vertices of the element at $1/3$ per vertex.

4.3.3.2 Definitions of the Variables

The Cartesian coordinates of the three nodes (1, 2, 3) of the triangular element (cf. Fig. 4.12) follow:

$$\mathbf{1} = \begin{vmatrix} x_1 \\ y_1 \\ z_1 \end{vmatrix} \tag{4.118}$$

$$\mathbf{2} = \begin{vmatrix} x_2 \\ y_2 \\ z_2 \end{vmatrix} \tag{4.119}$$

$$\mathbf{3} = \begin{vmatrix} x_3 \\ y_3 \\ z_3 \end{vmatrix} \tag{4.120}$$

Fig. 4.12 Example of triangular element. The drag forces are calculated for U twines and for V twines. The twine coordinates are noted in this example

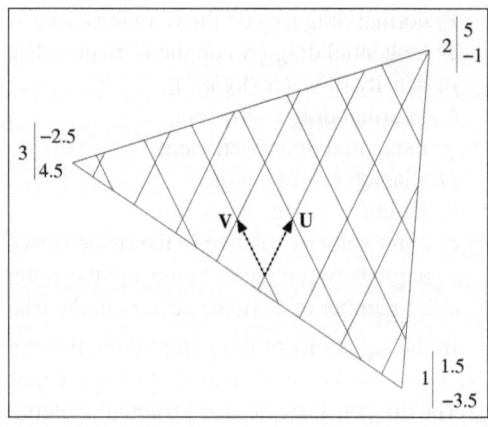

The twine coordinates of the three nodes (1, 2, 3) of the triangular element are as follows:

$$1 = \begin{vmatrix} U_1 \\ V_1 \end{vmatrix} \tag{4.121}$$

$$2 = \begin{vmatrix} U_2 \\ V_2 \end{vmatrix} \tag{4.122}$$

$$3 = \begin{vmatrix} U_3 \\ V_3 \end{vmatrix} \tag{4.123}$$

The vector current is

$$\mathbf{c} = \begin{vmatrix} c_x \\ c_y \\ c_z \end{vmatrix} \tag{4.124}$$

Generally speaking, c_z is null.
It has been seen previously:

$$\mathbf{U} = \frac{V_3 - V_1}{d} \mathbf{12} - \frac{V_2 - V_1}{d} \mathbf{13} \tag{4.125}$$

$$\mathbf{V} = \frac{U_2 - U_1}{d} \mathbf{13} - \frac{U_3 - U_1}{d} \mathbf{12} \tag{4.126}$$

with sides vectors:

$$\mathbf{12} = \begin{vmatrix} x_2 - x_1 \\ y_2 - y_1 \\ z_2 - z_1 \end{vmatrix} \tag{4.127}$$

$$\mathbf{13} = \begin{vmatrix} x_3 - x_1 \\ y_3 - y_1 \\ z_3 - z_1 \end{vmatrix} \tag{4.128}$$

and

$$d = (U_2 - U_1)(V_3 - V_1) - (U_3 - U_1)(V_2 - V_1) \tag{4.129}$$

The components of U twine vectors are as follows:

$$\mathbf{U} = \begin{vmatrix} U_x \\ U_y \\ U_z \end{vmatrix} \tag{4.130}$$

$$\mathbf{U} = \begin{vmatrix} \frac{1}{d} [(V_3 - V_1)(x_2 - x_1) - (V_2 - V_1)(x_3 - x_1)] \\ \frac{1}{d} [(V_3 - V_1)(y_2 - y_1) - (V_2 - V_1)(y_3 - y_1)] \\ \frac{1}{d} [(V_3 - V_1)(z_2 - z_1) - (V_2 - V_1)(z_3 - z_1)] \end{vmatrix} \tag{4.131}$$

The angle between current and U is

$$cos(\alpha) = \frac{\mathbf{c}.\mathbf{U}}{|\mathbf{c}||\mathbf{U}|} \tag{4.132}$$

The components of V twine vectors are as follows:

$$\mathbf{V} = \begin{vmatrix} V_x \\ V_y \\ V_z \end{vmatrix} \tag{4.133}$$

$$\mathbf{V} = \begin{vmatrix} \frac{1}{q}[(U_2 - U_1)(x_3 - x_1) - (U_3 - U_1)(x_2 - x_1)] \\ \frac{1}{q}[(U_2 - U_1)(y_3 - y_1) - (U_3 - U_1)(y_2 - y_1)] \\ \frac{1}{q}[(U_2 - U_1)(z_3 - z_1) - (U_3 - U_1)(z_2 - z_1)] \end{vmatrix} \tag{4.134}$$

The angle between current and V is

$$cos(\beta) = \frac{\mathbf{c}.\mathbf{V}}{|\mathbf{c}||\mathbf{V}|} \tag{4.135}$$

4.3.3.3 Stiffness of the Normal Force on the U Twines

The normal force on U twines is

$$\mathbf{F} = |\mathbf{F}|\frac{\mathbf{U} \wedge (\mathbf{c} \wedge \mathbf{U})}{|\mathbf{U} \wedge (\mathbf{c} \wedge \mathbf{U})|} \tag{4.136}$$

That means that the x y and z components are as follows:

$$\mathbf{F}_x = |\mathbf{F}|\frac{\mathbf{E}_x}{|\mathbf{E}|} \tag{4.137}$$

$$\mathbf{F}_y = |\mathbf{F}|\frac{\mathbf{E}_y}{|\mathbf{E}|} \tag{4.138}$$

$$\mathbf{F}_z = |\mathbf{F}|\frac{\mathbf{E}_z}{|\mathbf{E}|} \tag{4.139}$$

With:

$$\mathbf{E} = \mathbf{U} \wedge (\mathbf{c} \wedge \mathbf{U}) \tag{4.140}$$

and

$$\mathbf{E} = \begin{vmatrix} E_x \\ E_y \\ E_z \end{vmatrix} \tag{4.141}$$

The x component of the derivative is

$$F'_x = |F|'\frac{E_x}{|E|} + |F|\frac{E'_x|E| - E_x|E|'}{|E|^2} \tag{4.142}$$

Which gives for the x y and z components:

$$F'_x = |F|'\frac{E_x}{|E|} + \frac{|F|}{|E|^2}\left\{E'_x|E| - \frac{E_x}{|E|}(E_x E'_x + E_y E'_y + E_z E'_z)\right\} \tag{4.143}$$

$$F'_y = |F|'\frac{E_y}{|E|} + \frac{|F|}{|E|^2}\left\{E'_y|E| - \frac{E_y}{|E|}(E_x E'_x + E_y E'_y + E_z E'_z)\right\} \tag{4.144}$$

$$F'_z = |F|'\frac{E_z}{|E|} + \frac{|F|}{|E|^2}\left\{E'_z|E| - \frac{E_z}{|E|}(E_x E'_x + E_y E'_y + E_z E'_z)\right\} \tag{4.145}$$

For this assessment the derivative of E is required:

$$E' = U' \wedge (c \wedge U) + U \wedge (c \wedge U') \tag{4.146}$$

This leads to:

$$E' = 2(U'.U)c - (U'.c)U - (U.c)U' \tag{4.147}$$

Which is:

$$E'_x = 2(U'.U)c_x - (U'.c)U_x - (U.c)U'_x \tag{4.148}$$

$$E'_y = 2(U'.U)c_y - (U'.c)U_y - (U.c)U'_y \tag{4.149}$$

$$E'_z = 2(U'.U)c_z - (U'.c)U_z - (U.c)U'_z \tag{4.150}$$

With:

$$U'.U = U_x U'_x + U_y U'_y + U_z U'_z \tag{4.151}$$

$$U'.c = c_x U'_x + c_y U'_y + c_z U'_z \tag{4.152}$$

$$U.c = U_x c_x + U_y c_y + U_z c_z \tag{4.153}$$

The derivative of the amplitude of the normal force is

$$|F|' = \frac{1}{2}\rho C_d Dl_0|c|^2 \left([sin(\alpha)]^2\right)' \frac{d}{2} \tag{4.154}$$

Which is

$$|F|' = \frac{d}{2}\rho C_d Dl_0|c|^2 cos(\alpha)sin(\alpha)\alpha' \tag{4.155}$$

The derivative of α is

$$\alpha' = \frac{-1}{\sqrt{1 - \left(\frac{\mathbf{c}.\mathbf{U}}{|\mathbf{c}||\mathbf{U}|}\right)^2}} \left[\frac{\mathbf{c}.\mathbf{U}}{|\mathbf{c}||\mathbf{U}|}\right]' \tag{4.156}$$

That gives

$$\alpha' = \frac{-1}{\sqrt{1 - \left(\frac{\mathbf{c}.\mathbf{U}}{|\mathbf{c}||\mathbf{U}|}\right)^2}} \left[\frac{\mathbf{c}}{|\mathbf{c}|} \cdot \left(\frac{\mathbf{U}}{|\mathbf{U}|}\right)'\right] \tag{4.157}$$

The derivative of the U twine direction is

$$\left(\frac{\mathbf{U}}{|\mathbf{U}|}\right)' = \frac{\mathbf{U}'|\mathbf{U}| - \mathbf{U}|\mathbf{U}|'}{|\mathbf{U}|^2} \tag{4.158}$$

That means that the derivative of α is

$$\alpha' = \frac{-1}{\sqrt{1 - \left(\frac{\mathbf{c}.\mathbf{U}}{|\mathbf{c}||\mathbf{U}|}\right)^2}} \left(\frac{\mathbf{c}}{|\mathbf{c}|}\right) \cdot \left(\frac{\mathbf{U}'|\mathbf{U}| - \mathbf{U}|\mathbf{U}|'}{|\mathbf{U}|^2}\right) \tag{4.159}$$

or

$$\alpha' = \frac{-1}{|\mathbf{U}|^2|\mathbf{c}|\sin\alpha} \left\{|\mathbf{U}| \left[c_x\mathbf{U}'_x + c_y\mathbf{U}'_y + c_z\mathbf{U}'_z\right] - (\mathbf{c}.\mathbf{U})|\mathbf{U}|'\right\} \tag{4.160}$$

In this case $\mathbf{U}'_x$ is the component along x of $\mathbf{U}'$.
The derivative of vector $\mathbf{U}$ is

$$\mathbf{U}' = \begin{vmatrix} \mathbf{U}'_x \\ \mathbf{U}'_y \\ \mathbf{U}'_z \end{vmatrix} \tag{4.161}$$

Which is

$$\frac{\partial U_x}{\partial x_1} = \frac{\partial U_y}{\partial y_1} = \frac{\partial U_z}{\partial z_1} = \frac{1}{d}(V_2 - V_3) \tag{4.162}$$

$$\frac{\partial U_x}{\partial x_2} = \frac{\partial U_y}{\partial y_2} = \frac{\partial U_z}{\partial z_2} = \frac{1}{d}(V_3 - V_1) \tag{4.163}$$

$$\frac{\partial U_x}{\partial x_3} = \frac{\partial U_y}{\partial y_3} = \frac{\partial U_z}{\partial z_3} = \frac{1}{d}(V_1 - V_2) \tag{4.164}$$

$$\frac{\partial U_x}{\partial y_1} = \frac{\partial U_x}{\partial y_2} = \frac{\partial U_x}{\partial y_3} = \frac{\partial U_x}{\partial z_1} = \frac{\partial U_x}{\partial z_2} = \frac{\partial U_x}{\partial z_3} = 0 \qquad (4.165)$$

$$\frac{\partial U_y}{\partial z_1} = \frac{\partial U_y}{\partial z_2} = \frac{\partial U_y}{\partial z_3} = \frac{\partial U_y}{\partial x_1} = \frac{\partial U_y}{\partial x_2} = \frac{\partial U_y}{\partial x_3} = 0 \qquad (4.166)$$

$$\frac{\partial U_z}{\partial x_1} = \frac{\partial U_z}{\partial x_2} = \frac{\partial U_z}{\partial x_3} = \frac{\partial U_z}{\partial y_1} = \frac{\partial U_z}{\partial y_2} = \frac{\partial U_z}{\partial y_3} = 0 \qquad (4.167)$$

On vector form and for the nine coordinates of the triangular element it is:

$$\frac{\partial \mathbf{U}}{\partial x_1} = \begin{vmatrix} \frac{V_2 - V_3}{d} \\ 0 \\ 0 \end{vmatrix} \qquad (4.168)$$

$$\frac{\partial \mathbf{U}}{\partial y_1} = \begin{vmatrix} 0 \\ \frac{V_2 - V_3}{d} \\ 0 \end{vmatrix} \qquad (4.169)$$

$$\frac{\partial \mathbf{U}}{\partial z_1} = \begin{vmatrix} 0 \\ 0 \\ \frac{V_2 - V_3}{d} \end{vmatrix} \qquad (4.170)$$

$$\frac{\partial \mathbf{U}}{\partial x_2} = \begin{vmatrix} \frac{V_3 - V_1}{d} \\ 0 \\ 0 \end{vmatrix} \qquad (4.171)$$

$$\frac{\partial \mathbf{U}}{\partial y_2} = \begin{vmatrix} 0 \\ \frac{V_3 - V_1}{d} \\ 0 \end{vmatrix} \qquad (4.172)$$

$$\frac{\partial \mathbf{U}}{\partial z_2} = \begin{vmatrix} 0 \\ 0 \\ \frac{V_3 - V_1}{d} \end{vmatrix} \qquad (4.173)$$

$$\frac{\partial \mathbf{U}}{\partial x_3} = \begin{vmatrix} \frac{V_1 - V_2}{d} \\ 0 \\ 0 \end{vmatrix} \qquad (4.174)$$

$$\frac{\partial \mathbf{U}}{\partial y_3} = \begin{vmatrix} 0 \\ \frac{V_1 - V_2}{d} \\ 0 \end{vmatrix} \qquad (4.175)$$

$$\frac{\partial \mathbf{U}}{\partial z_3} = \begin{vmatrix} 0 \\ 0 \\ \frac{V_1 - V_2}{d} \end{vmatrix} \qquad (4.176)$$

The derivative of the norm of vector $\mathbf{U}$ is

$$|\mathbf{U}|' = \frac{U_x U_x' + U_y U_y' + U_z U_z'}{|\mathbf{U}|} \qquad (4.177)$$

This gives for the nine coordinates of the triangular element:

$$\frac{\partial |\mathbf{U}|}{\partial x_1} = \frac{U_x(V_2 - V_3)}{d|\mathbf{U}|} \tag{4.178}$$

$$\frac{\partial |\mathbf{U}|}{\partial y_1} = \frac{U_y(V_2 - V_3)}{d|\mathbf{U}|} \tag{4.179}$$

$$\frac{\partial |\mathbf{U}|}{\partial z_1} = \frac{U_z(V_2 - V_3)}{d|\mathbf{U}|} \tag{4.180}$$

$$\frac{\partial |\mathbf{U}|}{\partial x_2} = \frac{U_x(V_3 - V_1)}{d|\mathbf{U}|} \tag{4.181}$$

$$\frac{\partial |\mathbf{U}|}{\partial y_2} = \frac{U_y(V_3 - V_1)}{d|\mathbf{U}|} \tag{4.182}$$

$$\frac{\partial |\mathbf{U}|}{\partial z_2} = \frac{U_z(V_3 - V_1)}{d|\mathbf{U}|} \tag{4.183}$$

$$\frac{\partial |\mathbf{U}|}{\partial x_3} = \frac{U_x(V_1 - V_2)}{d|\mathbf{U}|} \tag{4.184}$$

$$\frac{\partial |\mathbf{U}|}{\partial y_3} = \frac{U_y(V_1 - V_2)}{d|\mathbf{U}|} \tag{4.185}$$

$$\frac{\partial |\mathbf{U}|}{\partial z_3} = \frac{U_z(V_1 - V_2)}{d|\mathbf{U}|} \tag{4.186}$$

This leads to the derivatives of α (angle between c and U):

$$\frac{\partial \alpha}{\partial x_1} = \frac{V_3 - V_2}{d|\mathbf{U}|^2|\mathbf{c}|\sqrt{1 - \left(\frac{\mathbf{c}.\mathbf{U}}{|\mathbf{c}||\mathbf{U}|}\right)^2}} \left[c_x|\mathbf{U}| - \frac{U_x}{|\mathbf{U}|}\mathbf{c}.\mathbf{U}\right] \tag{4.187}$$

$$\frac{\partial \alpha}{\partial y_1} = \frac{V_3 - V_2}{d|\mathbf{U}|^2|\mathbf{c}|\sqrt{1 - \left(\frac{\mathbf{c}.\mathbf{U}}{|\mathbf{c}||\mathbf{U}|}\right)^2}} \left[c_y|\mathbf{U}| - \frac{U_y}{|\mathbf{U}|}\mathbf{c}.\mathbf{U}\right] \tag{4.188}$$

$$\frac{\partial \alpha}{\partial z_1} = \frac{V_3 - V_2}{d|\mathbf{U}|^2|\mathbf{c}|\sqrt{1 - \left(\frac{\mathbf{c}.\mathbf{U}}{|\mathbf{c}||\mathbf{U}|}\right)^2}} \left[c_z|\mathbf{U}| - \frac{U_z}{|\mathbf{U}|}\mathbf{c}.\mathbf{U}\right] \tag{4.189}$$

$$\frac{\partial \alpha}{\partial x_2} = \frac{V_1 - V_3}{d|\mathbf{U}|^2|\mathbf{c}|\sqrt{1 - \left(\frac{\mathbf{c}.\mathbf{U}}{|\mathbf{c}||\mathbf{U}|}\right)^2}} \left[c_x|\mathbf{U}| - \frac{U_x}{|\mathbf{U}|}\mathbf{c}.\mathbf{U}\right] \tag{4.190}$$

$$\frac{\partial \alpha}{\partial y_2} = \frac{V_1 - V_3}{d|\mathbf{U}|^2|\mathbf{c}|\sqrt{1 - \left(\frac{\mathbf{c}.\mathbf{U}}{|\mathbf{c}||\mathbf{U}|}\right)^2}} \left[c_y|\mathbf{U}| - \frac{U_y}{|\mathbf{U}|}\mathbf{c}.\mathbf{U}\right] \tag{4.191}$$

$$\frac{\partial \alpha}{\partial z_2} = \frac{V_1 - V_3}{d|\mathbf{U}|^2|\mathbf{c}|\sqrt{1 - \left(\frac{\mathbf{c}.\mathbf{U}}{|\mathbf{c}||\mathbf{U}|}\right)^2}} \left[c_z|\mathbf{U}| - \frac{U_z}{|\mathbf{U}|}\mathbf{c}.\mathbf{U} \right] \qquad (4.192)$$

$$\frac{\partial \alpha}{\partial x_3} = \frac{V_2 - V_1}{d|\mathbf{U}|^2|\mathbf{c}|\sqrt{1 - \left(\frac{\mathbf{c}.\mathbf{U}}{|\mathbf{c}||\mathbf{U}|}\right)^2}} \left[c_x|\mathbf{U}| - \frac{U_x}{|\mathbf{U}|}\mathbf{c}.\mathbf{U} \right] \qquad (4.193)$$

$$\frac{\partial \alpha}{\partial y_3} = \frac{V_2 - V_1}{d|\mathbf{U}|^2|\mathbf{c}|\sqrt{1 - \left(\frac{\mathbf{c}.\mathbf{U}}{|\mathbf{c}||\mathbf{U}|}\right)^2}} \left[c_y|\mathbf{U}| - \frac{U_y}{|\mathbf{U}|}\mathbf{c}.\mathbf{U} \right] \qquad (4.194)$$

$$\frac{\partial \alpha}{\partial z_3} = \frac{V_2 - V_1}{d|\mathbf{U}|^2|\mathbf{c}|\sqrt{1 - \left(\frac{\mathbf{c}.\mathbf{U}}{|\mathbf{c}||\mathbf{U}|}\right)^2}} \left[c_z|\mathbf{U}| - \frac{U_z}{|\mathbf{U}|}\mathbf{c}.\mathbf{U} \right] \qquad (4.195)$$

4.3.3.4 Stiffness of the Tangential Force on the U Twines

The tangential force on U twines is

$$\mathbf{T} = |\mathbf{T}|\frac{\mathbf{F} \wedge (\mathbf{c} \wedge \mathbf{F})}{|\mathbf{F} \wedge (\mathbf{c} \wedge \mathbf{F})|} \qquad (4.196)$$

Following the definition of $\mathbf{F}_1$:

$$\mathbf{T} = |\mathbf{T}|\frac{[\mathbf{U} \wedge (\mathbf{c} \wedge \mathbf{U})] \wedge \{\mathbf{c} \wedge [\mathbf{U} \wedge (\mathbf{c} \wedge \mathbf{U})]\}}{|\,[\mathbf{U} \wedge (\mathbf{c} \wedge \mathbf{U})] \wedge \{\mathbf{c} \wedge [\mathbf{U} \wedge (\mathbf{c} \wedge \mathbf{U})]\}\,|} \qquad (4.197)$$

It follows that

$$\mathbf{T} = |\mathbf{T}|\frac{[(\mathbf{U}.\mathbf{U})(\mathbf{c}.\mathbf{c}) - (\mathbf{U}.\mathbf{c})^2](\mathbf{U}.\mathbf{c})\mathbf{U}}{|[(\mathbf{U}.\mathbf{U})(\mathbf{c}.\mathbf{c}) - (\mathbf{U}.\mathbf{c})^2](\mathbf{U}.\mathbf{c})\mathbf{U}|} \qquad (4.198)$$

or

$$\mathbf{T} = |\mathbf{T}|\frac{[|\mathbf{U}|^2|\mathbf{c}|^2 - (|\mathbf{U}||\mathbf{c}|cos\alpha)^2]|\mathbf{U}||\mathbf{c}|cos\alpha\mathbf{U}}{|[|\mathbf{U}|^2|\mathbf{c}|^2 - (|\mathbf{U}||\mathbf{c}|cos\alpha)^2]|\mathbf{U}||\mathbf{c}|cos\alpha\mathbf{U}|} \qquad (4.199)$$

and

$$\mathbf{T} = |\mathbf{T}|\frac{\cos\alpha\mathbf{U}}{|\cos\alpha||\mathbf{U}|} \qquad (4.200)$$

The x y and z components are as follows:

$$T_x = |\mathbf{T}|\frac{\cos\alpha U_x}{|\cos\alpha||\mathbf{U}|} \qquad (4.201)$$

$$T_y = |\mathbf{T}|\frac{\cos\alpha U_y}{|\cos\alpha||\mathbf{U}|} \qquad (4.202)$$

$$T_z = |\mathbf{T}| \frac{\cos \alpha \mathbf{U}_z}{|\cos \alpha||\mathbf{U}|} \tag{4.203}$$

The derivative of $\mathbf{T}_x$ is:

$$\mathbf{T}'_x = |\mathbf{T}|' \frac{\cos \alpha \mathbf{U}_x}{|\cos \alpha||\mathbf{U}|} + |\mathbf{T}| \frac{(\cos \alpha \mathbf{U}_x)'|\cos \alpha||\mathbf{U}| - \cos \alpha \mathbf{U}_x(|\cos \alpha||\mathbf{U}|)'}{(|\cos \alpha||\mathbf{U}|)^2} \tag{4.204}$$

$$\begin{aligned}
\mathbf{T}'_x = {} & |\mathbf{T}|' \frac{\cos \alpha \mathbf{U}_x}{|\cos \alpha||\mathbf{U}|} \\
& + \frac{|\mathbf{T}|}{|\cos \alpha||\mathbf{U}|} (\cos \alpha \mathbf{U}'_x - \sin \alpha \alpha' \mathbf{U}_x) \\
& - \frac{|\mathbf{T}| \cos \alpha \mathbf{U}_x}{(|\cos \alpha||\mathbf{U}|)^2} \left[|\cos \alpha| \frac{\mathbf{U}_x \mathbf{U}'_x + \mathbf{U}_y \mathbf{U}'_y + \mathbf{U}_z \mathbf{U}'_z}{|\mathbf{U}|} - \frac{\cos \alpha}{|\cos \alpha|} \sin \alpha \alpha' |\mathbf{U}| \right]
\end{aligned} \tag{4.205}$$

$$\begin{aligned}
\mathbf{T}'_x = {} & |\mathbf{T}|' \frac{\mathbf{T}_x}{|\mathbf{T}|} + \frac{|\mathbf{T}|}{|\cos \alpha||\mathbf{U}|} (\cos \alpha \mathbf{U}'_x - \sin \alpha \alpha' \mathbf{U}_x) \\
& - \frac{\mathbf{T}_x}{|\cos \alpha||\mathbf{U}|} \left[|\cos \alpha| \frac{\mathbf{U}_x \mathbf{U}'_x + \mathbf{U}_y \mathbf{U}'_y + \mathbf{U}_z \mathbf{U}'_z}{|\mathbf{U}|} - \frac{\cos \alpha}{|\cos \alpha|} \sin \alpha \alpha' |\mathbf{U}| \right]
\end{aligned} \tag{4.206}$$

$$\begin{aligned}
\mathbf{T}'_y = {} & |\mathbf{T}|' \frac{\mathbf{T}_y}{|\mathbf{T}|} + \frac{|\mathbf{T}|}{|\cos \alpha||\mathbf{U}|} (\cos \alpha \mathbf{U}'_y - \sin \alpha \alpha' \mathbf{U}_y) \\
& - \frac{\mathbf{T}_y}{|\cos \alpha||\mathbf{U}|} \left[|\cos \alpha| \frac{\mathbf{U}_x \mathbf{U}'_x + \mathbf{U}_y \mathbf{U}'_y + \mathbf{U}_z \mathbf{U}'_z}{|\mathbf{U}|} - \frac{\cos \alpha}{|\cos \alpha|} \sin \alpha \alpha' |\mathbf{U}| \right]
\end{aligned} \tag{4.207}$$

$$\begin{aligned}
\mathbf{T}'_z = {} & |\mathbf{T}|' \frac{\mathbf{T}_z}{|\mathbf{T}|} + \frac{|\mathbf{T}|}{|\cos \alpha||\mathbf{U}|} (\cos \alpha \mathbf{U}'_z - \sin \alpha \alpha' \mathbf{U}_z) \\
& - \frac{\mathbf{T}_z}{|\cos \alpha||\mathbf{U}|} \left[|\cos \alpha| \frac{\mathbf{U}_x \mathbf{U}'_x + \mathbf{U}_y \mathbf{U}'_y + \mathbf{U}_z \mathbf{U}'_z}{|\mathbf{U}|} - \frac{\cos \alpha}{|\cos \alpha|} \sin \alpha \alpha' |\mathbf{U}| \right]
\end{aligned} \tag{4.208}$$

The derivative of the amplitude of the tangential force is

$$|\mathbf{T}|' = f \frac{1}{2} \rho C_d Dl_0 |\mathbf{c}|^2 ([cos(\alpha)]^2)' \frac{d}{2} \tag{4.209}$$

which is

$$|\mathbf{T}|' = -\frac{d}{2} f \rho C_d Dl_0 |\mathbf{c}|^2 cos(\alpha) sin(\alpha) \alpha' \tag{4.210}$$

4.3.3.5 Stiffness of the Normal and Tangential Forces on the *V* Twines

This evaluations are identical to the previous, but with V and β used in place of U and α.

4.3.4 Twine Flexionin Netting Plane

The resistance to twine bending in the plane of the net is also called the mesh opening stiffness (Fig. 4.13). In a first approximation, this stiffness is neglected, but the use of steeper nets makes it necessary to take this mechanical phenomenon into account in numerical models. Currently, only [15, 12] and the present model take this mesh opening stiffness into account.

In the present model, the half angle (α) between the twine vectors (**U** and **V**) could lead to a couple between twine vectors (**U** and **V**). This angle is calculated by

$$\alpha = \frac{1}{2}acos\left(\frac{\mathbf{U}.\mathbf{V}}{|\mathbf{U}||\mathbf{V}|}\right) \tag{4.211}$$

The couple on a knot due to the U twine is equilibrated by the couple of the V twine; otherwise the knot would not be in equilibrium. These couples are approximated in the model by

$$C_u = -C_v = H(\alpha - \alpha_0) \tag{4.212}$$

Fig. 4.13 Demonstration of mesh opening stiffness. Deformation remains limited despite the weight added to the bottom of the net on (**b**)

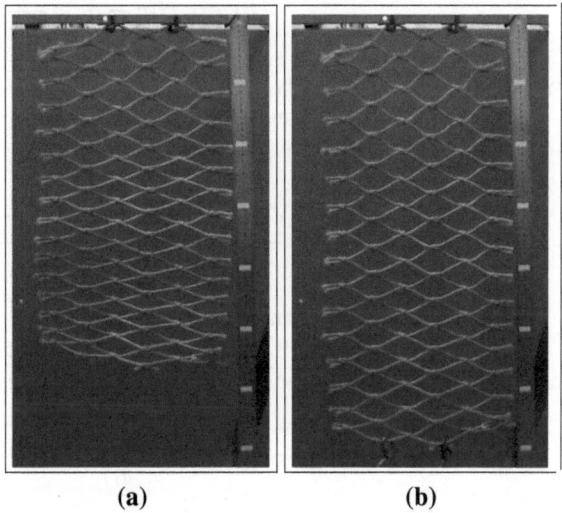

(**a**)   (**b**)

where α_0 is the angle between the unstressed twines (without couple on twines) and H is the mesh opening stiffness (N.m/Rad).

This couple varies linearly with the angle. [12, 15] suggest another formulation, since the twines are modelled as beams.

Forces at the vertices of the triangular element, mechanically equivalent to themesh opening stiffness, are calculated using the principle of virtual work:

If ∂x_1 is a virtual displacement along the x axis of vertex 1, then the external work (W_e) is

$$W_e = Fx_1 \partial x_1 \tag{4.213}$$

where Fx_1 is the effort along the x axis at vertex 1 of a triangular element.

This displacement creates a change in angle α, and therefore an internal work (W_i):

$$W_i = \frac{d}{2}(C_u \partial \alpha + C_v \partial \alpha) \tag{4.214}$$

$$d = (U_2 - U_1)(V_1 - V_3) - (U_3 - U_1)(V_1 - V_2) \tag{4.215}$$

where $d/2$ is the number of nodes in a triangular element.

Since the internal work is equal to the external work,

$$Fx_1 = C_u d \frac{\partial \alpha}{\partial x_1} \tag{4.216}$$

This gives, for all the force components at the vertices of the triangular element,

$$Fw_i = H(\alpha - \alpha_0)d \frac{\partial \alpha}{\partial w_i} \tag{4.217}$$

where $w = x$, y, and z, and $i = 1$, 2, and 3.

The derivative $\frac{\partial \alpha}{\partial w_i}$ of α relative to the coordinates w_i of vertices, which is necessary for calculating the forces, is

$$\frac{\partial \alpha}{\partial w_i} = \frac{\mathbf{V}_w v_i - \mathbf{U}_w u_i - \frac{\mathbf{U}_w(\mathbf{U.V})v_i}{|\mathbf{U}|^2} - \frac{\mathbf{V}_w(\mathbf{U.V})u_i}{|\mathbf{V}|^2}}{2d \sin(\alpha)|\mathbf{U}||\mathbf{V}|} \tag{4.218}$$

where $w = x$, y, and z, and $i = 1$, 2, and 3.

The stiffness matrix $(-\mathbf{F}'(\mathbf{X}))$ is completed by calculating the derivative component of efforts related to the coordinates of the vertices of the triangular element:

$$- \frac{\partial F_w i}{\partial t j} \tag{4.219}$$

where as above, $w = x$, y, and z, and $i = 1$, 2, and 3, and $t = x$, y, and z, and $j = 1$, 2, and 3.

Fig. 4.14 The net bends under its own weight, which highlights the bending stiffness of the net

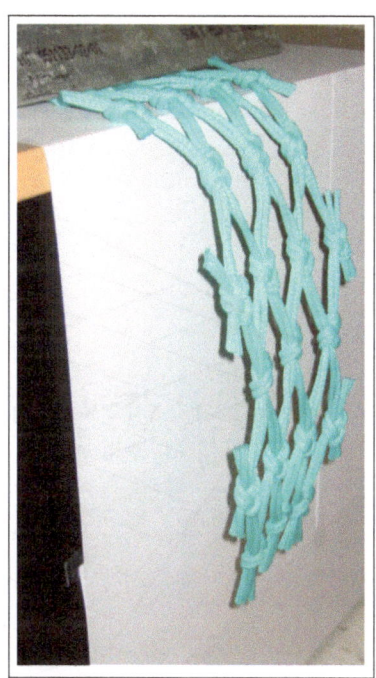

4.3.5 Twine Flexion Outside the Netting Plane

To our knowledge, no numerical model, except the present one, takes into account this mechanical property of the nets (Fig. 4.14). The angle between the U twine of a triangle ($\mathbf{U}_a$ in Fig. 4.15) and its neighbour ($\mathbf{U}_b$) is constant along the side common to the two triangular elements. This angle quantifies the bending of the twine.

The bending stiffness of the U twine tends to keep the twine straight. The equation governing the bending is as follows:

$$C = \frac{EI}{\rho} \qquad (4.220)$$

C: bending couple on the U twine (Nm),
EI: flexural stiffness, which is Young's modulus by inertia (Nm^2),
ρ: radius of curvature of the U twine (m).

This couple is generated, in the present modelling, when two successive triangular elements are bent or, more precisely, when the U twine is bent to the passage of a triangular element with its neighbour. The couple will then generate forces on the vertices (1, 2, 3, 4 in Fig. 4.15) on the two adjacent triangular elements. Obviously the bending of the V twines also leads to a couple. In the following only the effect

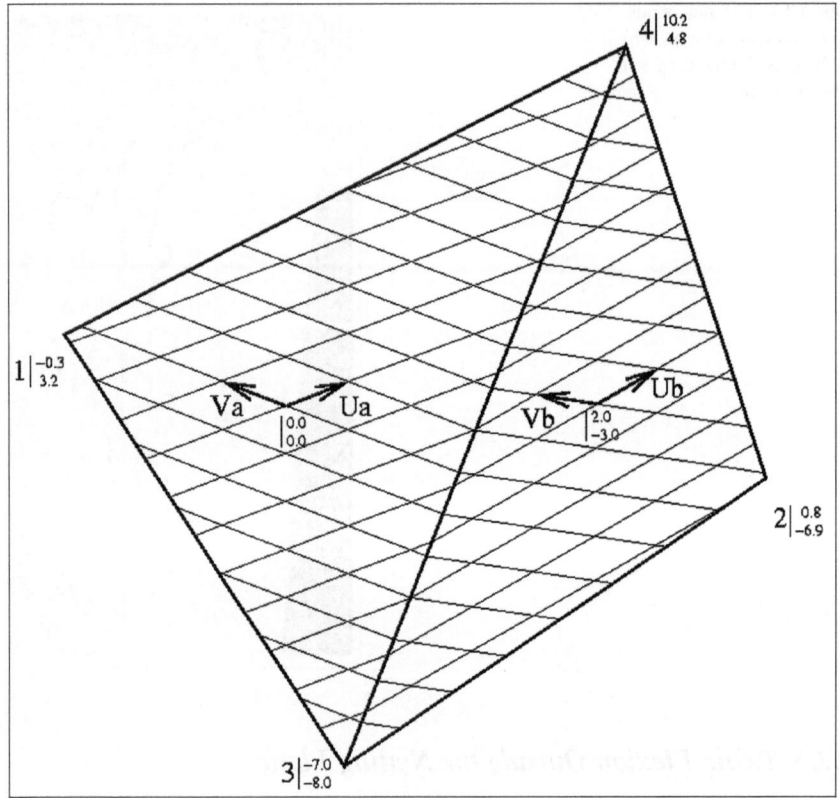

Fig. 4.15 Two triangular elements (134 and 243), the coordinates of which, in number of twines, are noted. The angle between the twine vectors $\mathbf{U_a}$ and $\mathbf{U_b}$ leads to a bending couple between the two triangular elements

of bending on the U twines is described; the bending on V twines has to be taken into account in the same way.

The radius of the curvature is estimated from the average lengths of twine U in each triangular element (Fig. 4.16). These average lengths are calculated using the average number of twine vectors ($\mathbf{U_a}$ and $\mathbf{U_b}$) by the U twine in the two triangular elements (n_a and n_b).

The twine vectors of the two triangular elements (see Sect. 4.2.1 p. 30) are as follows:

$$\mathbf{U}_a = \frac{V_4 - V_1}{d_a}\mathbf{13} - \frac{V_3 - V_1}{d_a}\mathbf{14} \qquad (4.221)$$

$$\mathbf{V}_a = \frac{U_4 - U_1}{d_a}\mathbf{13} - \frac{U_3 - U_1}{d_a}\mathbf{14} \qquad (4.222)$$

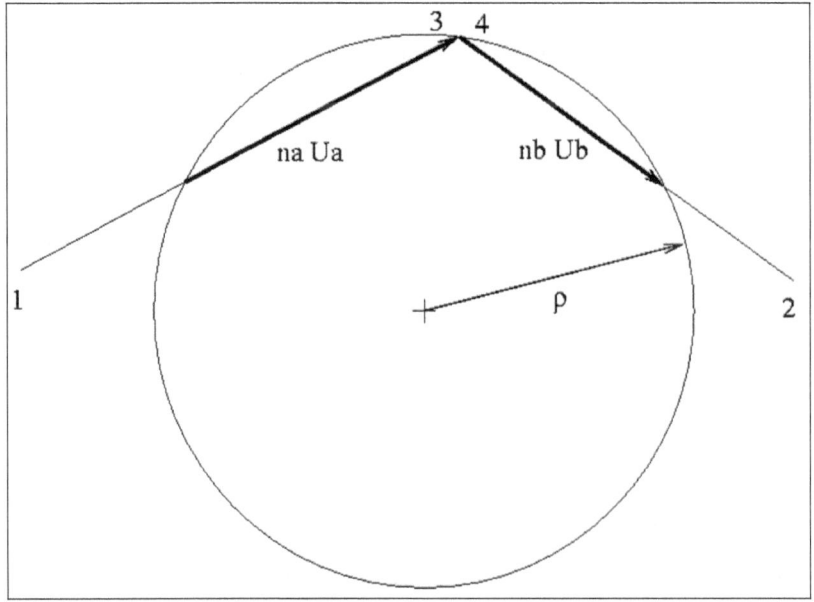

Fig. 4.16 Profile view of the two triangular elements. The radius of curvature (ρ) is estimated from the average length of twine vectors $\mathbf{U}$ in each triangle : $n_a \mathbf{U_a}$ and $n_b \mathbf{U_b}$

$$\mathbf{U}_b = \frac{V_3 - V_2}{d_b}\mathbf{24} - \frac{V_4 - V_2}{d_b}\mathbf{23} \tag{4.223}$$

$$\mathbf{V}_b = \frac{U_3 - U_2}{d_b}\mathbf{24} - \frac{U_4 - U_2}{d_b}\mathbf{23} \tag{4.224}$$

U_i, V_i: coordinates of vertex i in number of twines (twine coordinates). With side vectors:

$$\mathbf{13} = \begin{vmatrix} x_3 - x_1 \\ y_3 - y_1 \\ z_3 - z_1 \end{vmatrix} \tag{4.225}$$

$$\mathbf{24} = \begin{vmatrix} x_4 - x_2 \\ y_4 - y_2 \\ z_4 - z_2 \end{vmatrix} \tag{4.226}$$

The numbers of twine vectors ($\mathbf{U}_a$ and $\mathbf{U}_b$) for the U twines in the two triangular elements are

$$d_a = (U_3 - U_1)(V_4 - V_1) - (U_4 - U_1)(V_3 - V_1) \tag{4.227}$$

$$d_b = (U_4 - U_2)(V_3 - V_2) - (U_3 - U_2)(V_4 - V_2) \tag{4.228}$$

The average numbers of twine vectors ($\mathbf{U}_a$ and $\mathbf{U}_b$) by U twine are calculated from the number of twine vectors in the triangular elements and the length of the common side in twine coordinates ($V_3 - V_4$):

$$n_a = \frac{d_a}{2|V_3 - V_4|} \tag{4.229}$$

$$n_b = \frac{d_b}{2|V_3 - V_4|} \tag{4.230}$$

The radius of the curvature (ρ) is calculated from the circumscribed circle in the triangle of sides $na\mathbf{U}_a$, $nb\mathbf{U}_b$ and $na\mathbf{U}_a + nb\mathbf{U}_b$, as shown in Fig. 4.16. The side lengths of the triangle are

$$A = |n_a\mathbf{U}_a| \tag{4.231}$$

$$B = |n_b\mathbf{U}_b| \tag{4.232}$$

$$C = |n_a\mathbf{U}_a + n_b\mathbf{U}_b| \tag{4.233}$$

The equations of the triangle, which can be obtained in a mathematical compendium, give the radius of curvature:

$$\rho = \frac{ABC}{4S} \tag{4.234}$$

where S and p, the surface and the half perimeter of the triangle, are

$$S = \sqrt{p(p - A)(p - B)(p - C)} \tag{4.235}$$

$$p = \frac{A + B + C}{2} \tag{4.236}$$

The forces on the vertices (1, 2, 3 and 4) of the two triangular elements due to the twine bending are calculated using the principle of virtual work. In case of the X component of the force on vertex 1 (F_{x1}), a displacement ($\partial x1$) is defined along X axis of vertex 1. This displacement generates an external work:

$$W_e = F_{x1}\partial x1 \tag{4.237}$$

This movement also causes a variation of angle ($\partial\alpha$) between the twine vectors ($\mathbf{U}_a$ and $\mathbf{U}_b$) of the two triangular elements. This variation induces an internal work:

$$W_i = C\partial\alpha(V_3 - V_4) \tag{4.238}$$

According to the principle of virtual work, these works are equal, which gives the following:

$$F_{wi} = \frac{EI}{\rho}\frac{\partial\alpha}{\partial wi}(V_3 - V_4) \tag{4.239}$$

w: directions x, y, and z,
i: vertices 1, 2, 3, and 4,
$V_3 - V_4$: number of twines involved in the bending.

The angle α between the two twine vectors ($\mathbf{U}_a$ and $\mathbf{U}_b$) of the two triangular elements is calculated with the dot product of twine vectors (Fig. 4.16):

$$cos(\alpha) = \frac{\mathbf{U}_a.\mathbf{U}_b}{|\mathbf{U}_a||\mathbf{U}_b|} \tag{4.240}$$

The 12 derivatives of α relative to the coordinates of the vertices of the two triangular elements ($\frac{\partial \alpha}{\partial wi}$) are therefore required to calculate the effort on the vertices. They are as follows:

$$\frac{\partial \alpha}{\partial w1} = (V_3 - V_4)\frac{(\mathbf{U}_a.\mathbf{U}_b)U_{aw} - U_{bw}|\mathbf{U}_a|^2}{|\mathbf{U}_a|^3|\mathbf{U}_b|d_a sin(\alpha)} \tag{4.241}$$

$$\frac{\partial \alpha}{\partial w2} = (V_4 - V_3)\frac{(\mathbf{U}_a.\mathbf{U}_b)U_{bw} - U_{aw}|\mathbf{U}_b|^2}{|\mathbf{U}_b|^3|\mathbf{U}_a|d_b sin(\alpha)} \tag{4.242}$$

$$\frac{\partial \alpha}{\partial w3} = (V_4 - V_1)\frac{(\mathbf{U}_a.\mathbf{U}_b)U_{aw} - U_{bw}|\mathbf{U}_a|^2}{|\mathbf{U}_a|^3|\mathbf{U}_b|d_a sin(\alpha)} + (V_2 - V_4)\frac{(\mathbf{U}_a.\mathbf{U}_b)U_{bw} - U_{aw}|\mathbf{U}_b|^2}{|\mathbf{U}_b|^3|\mathbf{U}_a|d_b sin(\alpha)} \tag{4.243}$$

$$\frac{\partial \alpha}{\partial w4} = (V_1 - V_3)\frac{(\mathbf{U}_a.\mathbf{U}_b)U_{aw} - U_{bw}|\mathbf{U}_a|^2}{|\mathbf{U}_a|^3|\mathbf{U}_b|d_a sin(\alpha)} + (V_3 - V_2)\frac{(\mathbf{U}_a.\mathbf{U}_b)U_{bw} - U_{aw}|\mathbf{U}_b|^2}{|\mathbf{U}_b|^3|\mathbf{U}_a|d_b sin(\alpha)} \tag{4.244}$$

Here, U_{aw} is the component along the w axis of $\mathbf{U}_a$. In this case w is the axis consisting of x, y, and z. Obviously, U_{bw} is the component along the w axis of $\mathbf{U}_b$.

The efforts on the four vertices of the two triangular elements due to the bending of the U twine between these two elements have been previously calculated.

The stiffness matrix $(-F'(X))$ is completed by calculating the derivative of the 12 components of the forces relative to the 12 coordinates of the vertices of the two triangular elements. The 144 components of this matrix are

$$-\frac{\partial F_{wi}}{\partial tj} \tag{4.245}$$

With, as above:
w: x, y, and z.
i: 1, 2, 3, and 4.

And more:
t: x, y, and z,
j: 1, 2, 3, and 4.

4.3.6 Fish Catch Pressure

The mechanical effect of caught fish (Fig. 4.17) in a net is estimated by a pressure [1]. This pressure is exerted directly on the triangular elements in contact with the fish. In the case of water speed relative to that catch:

$$p = \frac{1}{2}\rho C_d v^2 \tag{4.246}$$

p: pressure of the catch on the net (Pa),
ρ: density of water (kg/m^3),
C_d: drag coefficient,
v: current amplitude (m/s).

This pressure is then applied to the surface of the triangular element $\left(\frac{12 \wedge 13}{2}\right)$. The resultant force is directed perpendicular to the triangular element. The effort on each vertex is that force by $1/3$.

$$\mathbf{F}_1 = \frac{12 \wedge 13}{2}\frac{p}{3} \tag{4.247}$$

$$\mathbf{F}_2 = \frac{12 \wedge 13}{2}\frac{p}{3} \tag{4.248}$$

$$\mathbf{F}_3 = \frac{12 \wedge 13}{2}\frac{p}{3} \tag{4.249}$$

With sides vectors:

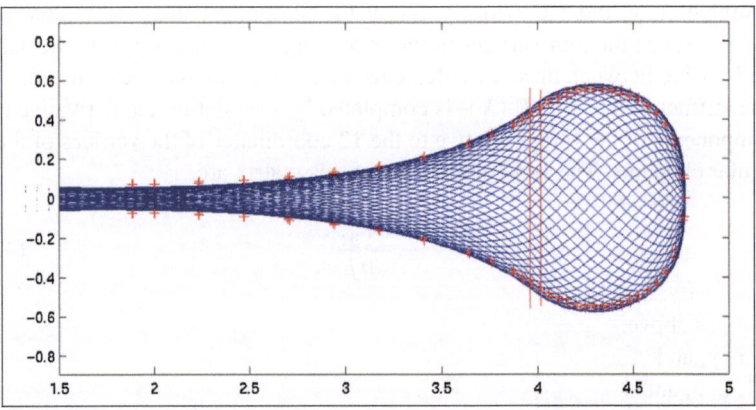

Fig. 4.17 Measurement in a flume tank tests (*cross*) and numerical modelling (*mesh*) for a scale (1/3) model of North Sea cod-end with 300 kg of catch

$$12 = \begin{vmatrix} x_2 - x_1 \\ y_2 - y_1 \\ z_2 - z_1 \end{vmatrix} \tag{4.250}$$

$$13 = \begin{vmatrix} x_3 - x_1 \\ y_3 - y_1 \\ z_3 - z_1 \end{vmatrix} \tag{4.251}$$

That gives:

$$\mathbf{F}_{1x} = \frac{p}{6}\left[(y_2 - y_1)(z_3 - z_1) - (z_2 - z_1)(y_3 - y_1)\right] \tag{4.252}$$

$$\mathbf{F}_{1y} = \frac{p}{6}\left[(z_2 - z_1)(x_3 - x_1) - (x_2 - x_1)(z_3 - z_1)\right] \tag{4.253}$$

$$\mathbf{F}_{1z} = \frac{p}{6}\left[(x_2 - x_1)(y_3 - y_1) - (y_2 - y_1)(x_3 - x_1)\right] \tag{4.254}$$

The contribution of this effect to the stiffness matrix is calculated through the derivatives of the forces. The derivatives of $\mathbf{F}_1$ is

$$\mathbf{F}'_1 = (12' \wedge 13 + 12 \wedge 13')\frac{p}{6} \tag{4.255}$$

The derivatives of $\mathbf{F}_1$, $\mathbf{F}_2$, and $\mathbf{F}_3$ are identical:

$$\frac{\partial \mathbf{F}_1}{\partial x_1} = \frac{p}{6}\begin{vmatrix} 0 \\ z_3 - z_2 \\ y_2 - y_3 \end{vmatrix} \tag{4.256}$$

$$\frac{\partial \mathbf{F}_1}{\partial y_1} = \frac{p}{6}\begin{vmatrix} z_2 - z_3 \\ 0 \\ x_3 - x_2 \end{vmatrix} \tag{4.257}$$

$$\frac{\partial \mathbf{F}_1}{\partial z_1} = \frac{p}{6}\begin{vmatrix} y_3 - y_2 \\ x_2 - x_3 \\ 0 \end{vmatrix} \tag{4.258}$$

$$\frac{\partial \mathbf{F}_1}{\partial x_2} = \frac{p}{6}\begin{vmatrix} 0 \\ z_1 - z_3 \\ y_3 - y_1 \end{vmatrix} \tag{4.259}$$

$$\frac{\partial \mathbf{F}_1}{\partial y_2} = \frac{p}{6}\begin{vmatrix} z_3 - z_1 \\ 0 \\ x_1 - x_3 \end{vmatrix} \tag{4.260}$$

$$\frac{\partial \mathbf{F}_1}{\partial z_2} = \frac{p}{6}\begin{vmatrix} y_1 - y_3 \\ x_3 - x_1 \\ 0 \end{vmatrix} \tag{4.261}$$

$$\frac{\partial \mathbf{F}_1}{\partial x_3} = \frac{p}{6} \begin{vmatrix} 0 \\ z_2 - z_1 \\ y_1 - y_2 \end{vmatrix}$$

(4.262)

$$\frac{\partial \mathbf{F}_1}{\partial y_3} = \frac{p}{6} \begin{vmatrix} z_1 - z_2 \\ 0 \\ x_2 - x_1 \end{vmatrix}$$

(4.263)

$$\frac{\partial \mathbf{F}_1}{\partial z_3} = \frac{p}{6} \begin{vmatrix} y_2 - y_1 \\ x_1 - x_2 \\ 0 \end{vmatrix}$$

(4.264)

4.3.7 Dynamic: Force of Inertia

The force of inertia is related to accelerations of the net and of the water particles just around the net. The calculation is done for each triangular element in three parts, one for each vertex, since the acceleration is not constant over the entire surface of each triangular element. Under these conditions, the parameters are local parameters at each vertex, including the acceleration and the mass. The mass per vertex is considered the third of the total mass of netting of the triangular element.

The force of inertia on each vertex of a triangular element mesh is estimated by [7]:

$$\mathbf{F}_i = M_a(\gamma_h - \gamma) + \rho V \gamma_h - M\gamma$$

(4.265)

$\mathbf{F}_i$: inertial force on the vertex i (N),
M_a: added mass (kg) of 1/3 of the triangular element,
M: mass of 1/3 of the net (kg),
V: volume of 1/3 of the net (m^3),
ρ : density of water (kg/m^3),
γ: acceleration of the vertex (m/s^2),
γ_h: acceleration of the water around the vertex (m/s^2).

The vertex speed is calculated as follows:

$$\mathbf{v} = \frac{\mathbf{x}_1 - \mathbf{x}}{\Delta t}$$

(4.266)

The acceleration of the vertex is

$$\gamma = \frac{\mathbf{v}_1 - \mathbf{v}}{\Delta t}$$

(4.267)

which gives

$$\gamma = \frac{\mathbf{x}_2 - 2\mathbf{x}_1 + \mathbf{x}}{\Delta t^2}$$

(4.268)

In this case, the contribution to the stiffness matrix, from the derivative of this inertia, is calculated by

$$- F' = -\frac{\partial \mathbf{F}_i}{\partial \mathbf{x}} \tag{4.269}$$

which leads to

$$- F' = (M + M_a)\frac{\partial \gamma}{\partial \mathbf{x}} \tag{4.270}$$

and

$$- F' = \frac{M + M_a}{\Delta t^2} \tag{4.271}$$

With: $\mathbf{x}$: position at t (m),
$\mathbf{x}_1$: position at $t - \Delta t$ (m),
$\mathbf{x}_2$: position at $t - 2\Delta t$ (m),
F': derivative of the force of inertia relative to the position (N/m),
Δt: time step (s).

4.3.8 Dynamic: Drag Force

The drag is related to the net and the relative speed of water particles just around the net. The calculation is done for each triangular element in three parts, one for each vertex, since this speed is not constant over the entire surface of each triangular element. Under these conditions the local parameters at each vertex are the vertex speed and one third of the number of twine vectors for the triangular element. The calculation is done for twines U and V.

The formulation for the twine drag is based on the assumptions of Landweber and Richtmeyer, as described earlier (Sect. 4.3.3, p. 46). The drag on the U twines applied on vertex i of the triangular element takes into account $1/3$ of the number of U twine vectors in the triangular element. This drag is as follows:

$$|\mathbf{F}_i| = \frac{d}{6}\frac{1}{2}\rho C_d D l_o(|\mathbf{c}_i|sin(\theta))^2 \tag{4.272}$$

$$|\mathbf{T}_i| = \frac{d}{6}f\frac{1}{2}\rho C_d D l_o(|\mathbf{c}_i|cos(\theta))^2 \tag{4.273}$$

F_i: normal force to the twines (N) on vertex i, this expression coming from the assumptions of Landweber,
T_i: tangential force (N) on vertex i, from Richtmeyer's assumption,
ρ: density of water (kg/m^3),
Cd: normal drag coefficient,
f: tangential coefficient,
D: diameter of twines U (m),
l_o: length of twine vectors U (m),

c_i: amplitude of the relative velocity of the water at vertex i (m/s),
θ: angle between the twine vectors U and the relative velocity (radians),
$\frac{d}{6}$: one third of the number of twine vectors U in the triangular element.

The angle θ between the twine vector $\mathbf{U}$ and the relative velocity is calculated by

$$cos(\theta) = \frac{\mathbf{c}_i \mathbf{U}}{|\mathbf{c}_i||\mathbf{U}|} \tag{4.274}$$

The directions of the drag in case of twine vector $\mathbf{U}$ are as follows:

$$\frac{\mathbf{F}_i}{|\mathbf{F}_i|} = \frac{\mathbf{U}}{|\mathbf{U}|} \wedge \frac{\mathbf{c}_i \wedge \mathbf{U}}{|\mathbf{c}_i||\mathbf{U}|} \tag{4.275}$$

$$\frac{\mathbf{T}_i}{|\mathbf{T}_i|} = \frac{\mathbf{F}_i}{|\mathbf{F}_i|} \wedge \frac{\mathbf{c}_i \wedge \mathbf{U}}{|\mathbf{c}_i||\mathbf{U}|} \tag{4.276}$$

The drag amplitude on twines V is calculated following the same scheme.

4.3.9 Buoyancy and Weight

Buoyancy and weight are vertical forces (along the z axis, if it is the vertical axis). Their expression is summed in the following:

$$F_z = d\pi \frac{D^2}{4} l_0 (\rho_{netting} - \rho)g \tag{4.277}$$

F_z: weight of the net once immersed (N),
d: number of twine vectors $\mathbf{U}$ and twine vectors $\mathbf{V}$ per triangular element,
ρ: water density (kg/m^3),
$\rho_{netting}$: net density (kg/m^3),
D: diameter of twines (m),
g: gravity of the Earth (around 9.81 m/s^2),
l_0: length of twine vectors (m).

The length of the twine vectors is approximated by the unstretched twine vector l_0, since the elongation is generally quite small.

There is a contribution of this force to the stiffness matrix when the netting crosses the water surface. In this case there is a variation of force with the immersion. This contribution is not described here.

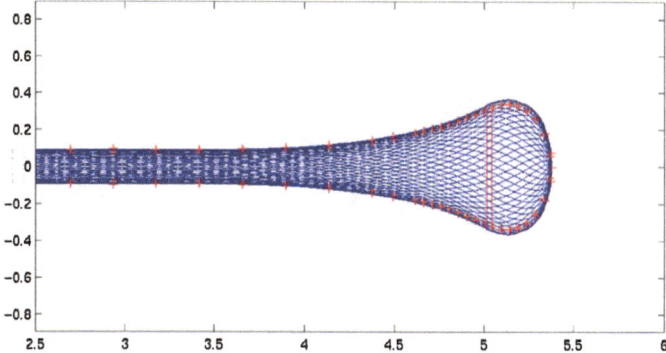

Fig. 4.18 Comparison between simulations (*net*) and flume tank tests (*crosses*) of trawl cod-ends [1]. Between 2.5 and 3.5 m the diameter is constant. This is due to contact between the nodes of the net

4.3.10 Contact Between Knots

It happens quite frequently that the nets are so close that the nodes come into contact with each other. This contact limits the closing of mesh (Fig. 4.18).

An effort similar to that described in Sect. 4.3.4 (p. 57) has been introduced to take into account this feature. This effort appears only when the twines are close enough, that is, when the angle between U and V twines is below a critical angle (α_{mini}). This angle is related to the node size and mesh side as follows (Fig. 4.19):

$$\alpha_{mini} = 2 \arcsin \left[\frac{knot_{size}}{2mesh_{side}} \right] \qquad (4.278)$$

α_{mini}: limit angle of contact between twines (rad),
$knot_{size}$: size of the node (m),
$mesh_{side}$: side of the mesh or length of twine vectors (m).

The $mesh_{side}$ could be the length of the twine vector along the U twine ($|U|$) or the length of the twine vector along the V twine ($|V|$). To avoid this choice (between $|U|$ and $|V|$), this length can be approximated by the unstretched length l_0 of the twine vector.

A couple is generated between the twines if the angle between them is less than the minimal angle:

$$\begin{cases} C = H(\alpha - \alpha_{mini}) \ if \ \alpha <= \alpha_{mini} \\ C = 0 \qquad\qquad\quad if \ \alpha > \alpha_{mini} \end{cases} \qquad (4.279)$$

C: couple between the twines due to the contact between knots (Nm),
α: angle between twines U and V (rad),

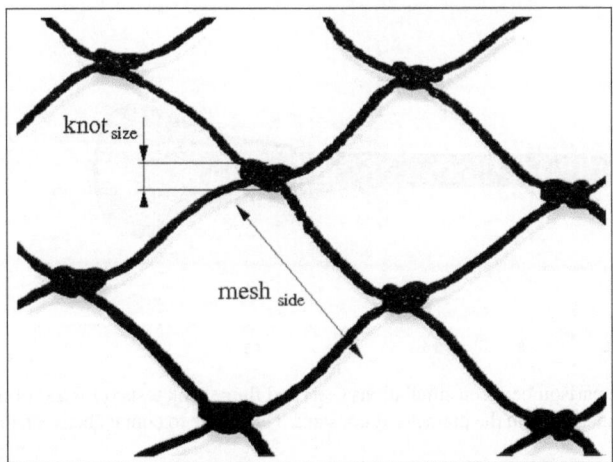

Fig. 4.19 The size of the knot limits the closure of the mesh. The minimal angle between twines is due to the size of the knot and the side of the mesh (which is also the length of twine vector)

H: stiffness (Nm/Rad).

This stiffness is not well known. Therefore, arbitrary values can be used, such as the following, proportional to the elongation stiffness of the twine (EA) (Fig. 4.19):

$$H = \frac{1}{100} \frac{mesh^2_{side} EA}{knot_{size}} \qquad (4.280)$$

A: section of the twine (m^2),
E: Young's modulus (Pa).

The forces on the vertices of triangular elements and the stiffness use the same expressions as those described in Sect. 4.3.4 (p. 57).

Chapter 5
The Bar Finite Element for Cable

Abstract The modelling for cable is described. A bar element for cable is described. The forces as well as the stiffness are given in case of tension in cables, of flexion and of hydrodynamic forces.

Keywords Bar element for cable · Tension in cables · Flexion of cables

5.1 Principle

The cables are split into bar elements (Fig. 5.1). The greater the number of bars, the better the representation of the curvature.

From the position $\mathbf{X}$ of the extremities of the bar elements the forces $\mathbf{F}$ on these extremities are calculated. The bar elements, in the present modelling, respect a couple of hypotheses. The first is that the bar element is straight. The second is that the bar element is elastic. These hypotheses make possible the calculation of forces on the extremities of the bar element.

5.2 Tension on Bars

5.2.1 Force Vector

The forces on the extremities of the bar elements are due to the tension in the bar (Fig. 5.2).

If the position of the extremities are noted 1 and 2, the length of the bar is:

$$l = \sqrt{\mathbf{12.12}} \tag{5.1}$$

D. Priour, *A Finite Element Method for Netting*, SpringerBriefs in Environmental Science, DOI: 10.1007/978-94-007-6844-4_5, © The Author(s) 2013

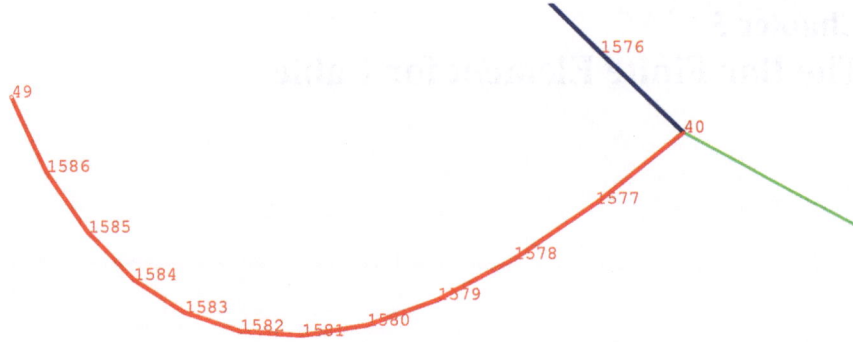

Fig. 5.1 View of three cables split into bar elements. The nodes number are noted

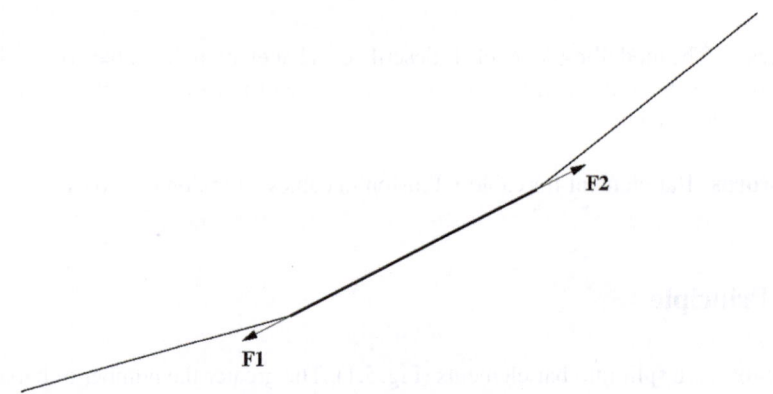

Fig. 5.2 Tension forces **F1** and **F2** on the extremities of the bar due to its tension

With:

$$\mathbf{12} = \begin{vmatrix} x_2 - x_1 \\ y_2 - y_1 \\ z_2 - z_1 \end{vmatrix} \tag{5.2}$$

The tension in the bar is:

$$|\mathbf{F}| = \frac{l - l_0}{l_0} EA \tag{5.3}$$

E : Young's modulus of the material (N/m^2),
A : mechanical section of the cable (m^2),
l_o : unstretched length of the bar element (m).

The force vectors on the two extremities of the bar are

$$\mathbf{F}_1 = |\mathbf{F}| \frac{\mathbf{21}}{l} \tag{5.4}$$

$$\mathbf{F}_2 = |\mathbf{F}| \frac{12}{l} \tag{5.5}$$

The components of these forces are:

$$
\begin{aligned}
\mathbf{F}_{1x} &= |\mathbf{F}| \frac{x_1 - x_2}{l} \\
\mathbf{F}_{1y} &= |\mathbf{F}| \frac{y_1 - y_2}{l} \\
\mathbf{F}_{1z} &= |\mathbf{F}| \frac{z_1 - z_2}{l} \\
\mathbf{F}_{2x} &= |\mathbf{F}| \frac{x_2 - x_1}{l} \\
\mathbf{F}_{2y} &= |\mathbf{F}| \frac{y_2 - y_1}{l} \\
\mathbf{F}_{2z} &= |\mathbf{F}| \frac{z_2 - z_1}{l}
\end{aligned}
\tag{5.6}
$$

5.2.2 Stiffness Matrix

The stiffness matrix is as follows:

$$
K = \begin{pmatrix}
-\dfrac{\partial F_{1x}}{\partial x_1} & -\dfrac{\partial F_{1x}}{\partial y_1} & -\dfrac{\partial F_{1x}}{\partial z_1} & -\dfrac{\partial F_{1x}}{\partial x_2} & -\dfrac{\partial F_{1x}}{\partial y_2} & -\dfrac{\partial F_{1x}}{\partial z_2} \\[2ex]
-\dfrac{\partial F_{1y}}{\partial x_1} & -\dfrac{\partial F_{1y}}{\partial y_1} & -\dfrac{\partial F_{1y}}{\partial z_1} & -\dfrac{\partial F_{1y}}{\partial x_2} & -\dfrac{\partial F_{1y}}{\partial y_2} & -\dfrac{\partial F_{1y}}{\partial z_2} \\[2ex]
-\dfrac{\partial F_{1z}}{\partial x_1} & -\dfrac{\partial F_{1z}}{\partial y_1} & -\dfrac{\partial F_{1z}}{\partial z_1} & -\dfrac{\partial F_{1z}}{\partial x_2} & -\dfrac{\partial F_{1z}}{\partial y_2} & -\dfrac{\partial F_{1z}}{\partial z_2} \\[2ex]
-\dfrac{\partial F_{2x}}{\partial x_1} & -\dfrac{\partial F_{2x}}{\partial y_1} & -\dfrac{\partial F_{2x}}{\partial z_1} & -\dfrac{\partial F_{2x}}{\partial x_2} & -\dfrac{\partial F_{2x}}{\partial y_2} & -\dfrac{\partial F_{2x}}{\partial z_2} \\[2ex]
-\dfrac{\partial F_{2y}}{\partial x_1} & -\dfrac{\partial F_{2y}}{\partial y_1} & -\dfrac{\partial F_{2y}}{\partial z_1} & -\dfrac{\partial F_{2y}}{\partial x_2} & -\dfrac{\partial F_{2y}}{\partial y_2} & -\dfrac{\partial F_{2y}}{\partial z_2} \\[2ex]
-\dfrac{\partial F_{2z}}{\partial x_1} & -\dfrac{\partial F_{2z}}{\partial y_1} & -\dfrac{\partial F_{2z}}{\partial z_1} & -\dfrac{\partial F_{2z}}{\partial x_2} & -\dfrac{\partial F_{2z}}{\partial y_2} & -\dfrac{\partial F_{2z}}{\partial z_2}
\end{pmatrix}
\tag{5.7}
$$

The stiffness matrix is calculated through the derivatives of force components. For the first component that gives:

$$
-\frac{\partial F_{1x}}{\partial x_1} = -\frac{\left[\frac{EA}{l_0} \frac{\partial l}{\partial x_1}(x_1 - x_2) + |\mathbf{F}| \frac{\partial (x_1 - x_2)}{\partial x_1} \right] l - |\mathbf{F}|(x_1 - x_2) \frac{\partial l}{\partial x_1}}{l^2}
\tag{5.8}
$$

with

$$
\frac{\partial l}{\partial x_1} = \frac{x_2 - x_1}{l}
\tag{5.9}
$$

That gives for the 36 components:

$$-\frac{\partial F_{1x}}{\partial x_1} = \frac{\partial F_{1x}}{\partial x_2} = \frac{\partial F_{2x}}{\partial x_1} = -\frac{\partial F_{2x}}{\partial x_2} = \frac{EA}{l^3 lo}\left[l^3 - l^2 lo + lo(x_2 - x_1)^2\right]$$

$$-\frac{\partial F_{1y}}{\partial y_1} = \frac{\partial F_{1y}}{\partial y_2} = \frac{\partial F_{2y}}{\partial y_1} = -\frac{\partial F_{2y}}{\partial y_2} = \frac{EA}{l^3 lo}\left[l^3 - l^2 lo + lo(y_2 - y_1)^2\right]$$

$$-\frac{\partial F_{1z}}{\partial z_1} = \frac{\partial F_{1z}}{\partial z_2} = \frac{\partial F_{2z}}{\partial z_1} = -\frac{\partial F_{2z}}{\partial z_2} = \frac{EA}{l^3 lo}\left[l^3 - l^2 lo + lo(z_2 - z_1)^2\right]$$

$$-\frac{\partial F_{1x}}{\partial y_1} = -\frac{\partial F_{1y}}{\partial x_1} = -\frac{\partial F_{2y}}{\partial x_2} = -\frac{\partial F_{2x}}{\partial y_2} = \frac{\partial F_{2y}}{\partial x_1} = \frac{\partial F_{2x}}{\partial y_1} = \frac{\partial F_{1y}}{\partial x_2} = \frac{\partial F_{1x}}{\partial y_2}$$

$$= \frac{EA}{l^3}\left[(x_2 - x_1)(y_2 - y_1)\right]$$

$$-\frac{\partial F_{1x}}{\partial z_1} = -\frac{\partial F_{1z}}{\partial x_1} = -\frac{\partial F_{2z}}{\partial x_2} = -\frac{\partial F_{2x}}{\partial z_2} = \frac{\partial F_{2z}}{\partial x_1} = \frac{\partial F_{2x}}{\partial z_1} = \frac{\partial F_{1z}}{\partial x_2} = \frac{\partial F_{1x}}{\partial z_2}$$

$$= \frac{EA}{l^3}\left[(x_2 - x_1)(z_2 - z_1)\right]$$

$$-\frac{\partial F_{1y}}{\partial z_1} = -\frac{\partial F_{1z}}{\partial y_1} = -\frac{\partial F_{2z}}{\partial y_2} = -\frac{\partial F_{2y}}{\partial z_2} = \frac{\partial F_{2z}}{\partial y_1} = \frac{\partial F_{2y}}{\partial z_1} = \frac{\partial F_{1z}}{\partial y_2} = \frac{\partial F_{1y}}{\partial z_2}$$

$$= \frac{EA}{l^3}\left[(y_2 - y_1)(z_2 - z_1)\right]$$

$$(5.10)$$

5.3 Bending of Cables

Cables could have a resistance in bending, such as beams. Beam deformation relates the curvature of the beam to the couple, such as:

$$C_o = \frac{EI}{R} \tag{5.11}$$

C_o: the couple on any point of the cable (N.m),
EI: the bending rigidity of the cable (N.m^2),
R: the radius of the cable at the point (m).

To take into account this behaviour in the numerical model, the cables are split into bar elements (Fig. 5.3). In case of bending stiffness, there is a couple C_o between consecutive bar elements (Fig. 5.4). This couple leads to forces on the extremities of theses two elements.

5.3.1 Force Vector

The forces on the extremities of two consecutive bar elements are due to the bending between the bar elements (Fig. 5.4).

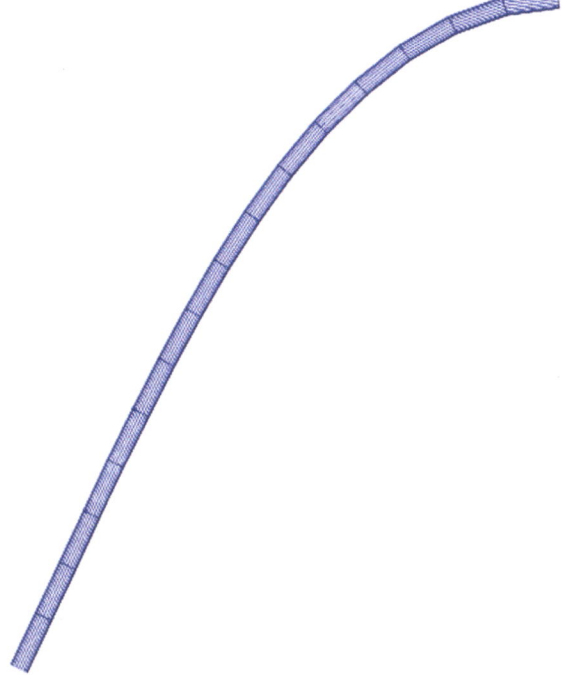

Fig. 5.3 The cable is embedded at *top right* and bends under its own weight. It is modelled with bar elements. Each bar is straight and articulated with its neighbour

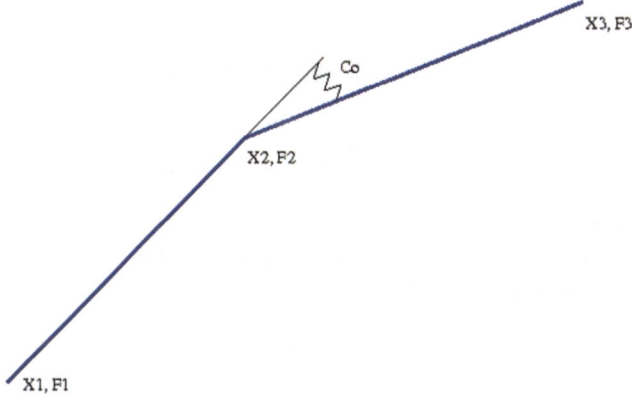

Fig. 5.4 Representation of two consecutive bars. A couple is introduced to take into account the bending rigidity of the cable. The spring symbolizes the couple

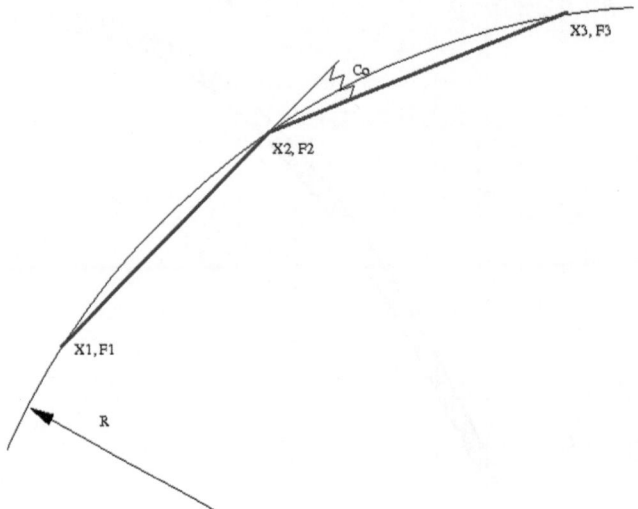

Fig. 5.5 The radius of the curvature is assessed by the *circle* passing by the extremities of the two bar elements

The curvature is approximated by the circle passing by the extremities of the two bar elements. The positions of the extremities of the bars allow assessment of this radius (Fig. 5.5). From this radius, and if the bending rigidity is known, the model is able to calculate the couple:

$$C_o = \frac{EI}{R} \tag{5.12}$$

The radius (R) is calculated from the position of the extremities:

$$R = \frac{ABC}{4\sqrt{p(p-A)(p-B)(p-C)}} \tag{5.13}$$

A (B): length of the first (second) bar (m),
C: distance between the extremities 1 and 3 in Fig. 5.5 (m),
p: the half perimeter (m), where

$$p = \frac{A+B+C}{2} \tag{5.14}$$

$$A = \begin{vmatrix} x_2 - x_1 \\ y_2 - y_1 \\ z_2 - z_1 \end{vmatrix} \tag{5.15}$$

$$B = \begin{vmatrix} x_3 - x_2 \\ y_3 - y_2 \\ z_3 - z_2 \end{vmatrix} \tag{5.16}$$

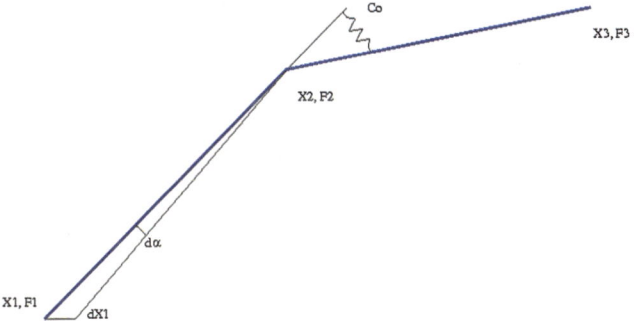

Fig. 5.6 A virtual displacement $(\partial x1)$ leads to an external work $(F_{x1}\partial x1)$ equal to the internal work $(C_o\partial\alpha)$

$$C = A + B$$
$$A = |\mathbf{A}|$$
$$B = |\mathbf{B}|$$
$$C = |\mathbf{A} + \mathbf{B}|$$

(5.17)

Once the couple C_o is calculated, the model assesses the forces on the extremities of the bars using the virtual work principle.

The force component along X on the extremity 1 of the first bar element is estimated by considering a virtual displacement $(\partial x1)$ along the axis x of the extremity 1 (Fig. 5.6). This displacement leads to an external work, considering $\partial x1$ small and consequently F_{x1} constant:

$$W_e = F_{x1}\partial x1$$

(5.18)

This virtual displacement also induces a change in the angle (α) between bar elements.

This virtual displacement leads to a variation of angle between bars $(\partial\alpha)$, and this variation of angle generates an internal work. If $\partial x1$ is small, $\partial\alpha$ is small and consequently C_o is constant. That gives

$$W_i = C_o\partial\alpha$$

(5.19)

Because the forces on the extremities of the two bar elements represent the couple C_o there is equality between the works. That leads to:

$$F_{x1} = C_o\frac{\partial\alpha}{\partial x1} \quad F_{x2} = C_o\frac{\partial\alpha}{\partial x2} \quad F_{x3} = C_o\frac{\partial\alpha}{\partial x3}$$
$$F_{y1} = C_o\frac{\partial\alpha}{\partial y1} \quad F_{y2} = C_o\frac{\partial\alpha}{\partial y2} \quad F_{y3} = C_o\frac{\partial\alpha}{\partial y3}$$
$$F_{z1} = C_o\frac{\partial\alpha}{\partial z1} \quad F_{z2} = C_o\frac{\partial\alpha}{\partial z2} \quad F_{z3} = C_o\frac{\partial\alpha}{\partial z3}$$

(5.20)

These forces components are:

$$F_{x1} = \frac{EI}{R \sin \alpha} \left[\frac{(x2 - x1)\mathbf{AB}}{A^3 B} + \frac{x2 - x3}{AB} \right]$$

$$F_{y1} = \frac{EI}{R \sin \alpha} \left[\frac{(y2 - y1)\mathbf{AB}}{A^3 B} + \frac{y2 - y3}{AB} \right] \qquad (5.21)$$

$$F_{z1} = \frac{EI}{R \sin \alpha} \left[\frac{(z2 - z1)\mathbf{AB}}{A^3 B} + \frac{z2 - z3}{AB} \right]$$

$$F_{x2} = \frac{EI}{R \sin \alpha} \left[\frac{(x1 - x2)\mathbf{AB}}{A^3 B} + \frac{(x3 - x2)\mathbf{AB}}{AB^3} + \frac{x3 - 2x2 + x1}{AB} \right]$$

$$F_{y2} = \frac{EI}{R \sin \alpha} \left[\frac{(y1 - y2)\mathbf{AB}}{A^3 B} + \frac{(y3 - y2)\mathbf{AB}}{AB^3} + \frac{y3 - 2y2 + y1}{AB} \right] \qquad (5.22)$$

$$F_{z2} = \frac{EI}{R \sin \alpha} \left[\frac{(z1 - z2)\mathbf{AB}}{A^3 B} + \frac{(z3 - z2)\mathbf{AB}}{AB^3} + \frac{z3 - 2z2 + z1}{AB} \right]$$

$$F_{x3} = \frac{EI}{R \sin \alpha} \left[\frac{(x2 - x3)\mathbf{AB}}{AB^3} + \frac{x2 - x1}{AB} \right]$$

$$F_{y3} = \frac{EI}{R \sin \alpha} \left[\frac{(y2 - y3)\mathbf{AB}}{AB^3} + \frac{y2 - y1}{AB} \right] \qquad (5.23)$$

$$F_{z3} = \frac{EI}{R \sin \alpha} \left[\frac{(z2 - z3)\mathbf{AB}}{AB^3} + \frac{z2 - z1}{AB} \right]$$

On vectorial form:

$$\mathbf{F}_1 = \frac{EI}{ABR \sin \alpha} \left[\frac{\mathbf{A}.\mathbf{AB}}{A^2} - \mathbf{B} \right]$$

$$\mathbf{F}_2 = \frac{EI}{ABR \sin \alpha} \left[-\frac{\mathbf{A}.\mathbf{AB}}{A^2} + \frac{\mathbf{B}.\mathbf{AB}}{B^2} + \mathbf{B} - \mathbf{A} \right] \qquad (5.24)$$

$$\mathbf{F}_3 = \frac{EI}{ABR \sin \alpha} \left[-\frac{\mathbf{B}.\mathbf{AB}}{B^2} + \mathbf{A} \right]$$

With:
$\mathbf{F}_1$ ($\mathbf{F}_2$, $\mathbf{F}_3$): force on the node 1 (2, 3),
$\mathbf{AB}$: scalar product between the two bar vectors,
$\mathbf{A}$ ($\mathbf{B}$): vector along the first (second) bar element,
A (B): length of the first (second) bar element (m),
$x1$ to $z3$: the Cartesian coordinates of the three extremities of the two bar elements (m).

5.3.2 Stiffness Matrix

The stiffness matrix is calculated with the derivatives of the force components (F_{x1} to F_{z3}) relative to the positions ($x1$ to $z3$). This means that the stiffness matrix has 81 components.

5.4 Drag on Cables

5.4.1 Introduction

The drag force on cables is calculated in this model as the contribution of the drag force on each bar elements. The formulation for the drag is based on the assumptions of Morrison, as adapted by Landweber and Richtmeyer (see Sect. 4.3.3, p. 46).

The drag amplitudes on bar element used in the model (Fig. 5.7) are

$$|\mathbf{F}| = \frac{1}{2}\rho C_d Dl_0 \, [|\mathbf{c}| \sin(\alpha)]^2$$
$$|\mathbf{T}| = f\frac{1}{2}\rho C_d Dl_0 \, [|\mathbf{c}| \cos(\alpha)]^2 \tag{5.25}$$

The directions of the drag are as follows:

$$\frac{\mathbf{F}}{|\mathbf{F}|} = \frac{\mathbf{B} \wedge (\mathbf{c} \wedge \mathbf{B})}{|\mathbf{B} \wedge (\mathbf{c} \wedge \mathbf{B})|}$$
$$\frac{\mathbf{T}}{|\mathbf{T}|} = \frac{\mathbf{F} \wedge (\mathbf{c} \wedge \mathbf{F})}{|\mathbf{F} \wedge (\mathbf{c} \wedge \mathbf{F})|} \tag{5.26}$$

$\mathbf{F}$: normal drag (N), following the assumptions of Landweber,
$\mathbf{T}$: tangential drag (N), Richtmeyer hypothesis,
$\mathbf{B}$: bar element vector,
ρ: density of water (kg/m^3),
C_d: normal drag coefficient,
f: tangential drag coefficient,
D: diameter of the bar element (m),
l_0: length of the bar element (m),
$\mathbf{c}$: water velocity relative to the bar element (m/s),
α: angle between the bar element and the water velocity (radians).

In the equations of drag amplitude, the expressions $|\mathbf{c}| \sin(\alpha)$ and $|\mathbf{c}| \cos(\alpha)$ are the normal and tangential projections on $\mathbf{c}$ along the bar element vector.

The length of the bar element used in the formulation of drag amplitude could be assessed by $|\mathbf{B}|$. That would mean it takes into account the bar element elongation.

Fig. 5.7 Normal (**F**) and
tangential (**T**) forces on a bar
element due to the velocity of
water (**c**)

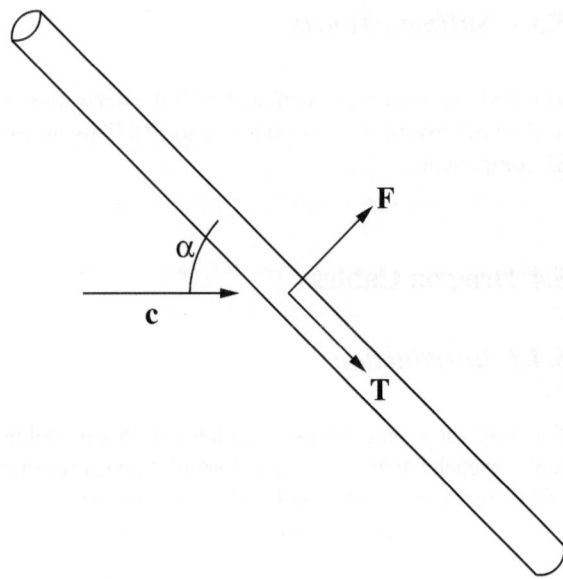

Generally speaking, a bar elongation is associated with a diameter D reduction by
the Poisson coefficient. Because this Poisson coefficient is not taken into account in
the present modelling, the bar element surface is approximated by Dl_0, where D is
the diameter of the bar and l_0 is the unstretched length of the bar element vectors.

All parameters, including the angle α are constant and known for each bar element.
Therefore, the drag can be calculated for each bar element. The drag force for a bar
element is spread over the two vertices of the element at $1/2$ per vertex.

5.4.2 Definitions of the Variables

The Cartesian coordinates of the two nodes $(1, 2)$ of the bar element are the following:

$$\mathbf{1} = \begin{vmatrix} x_1 \\ y_1 \\ z_1 \end{vmatrix}$$

$$\mathbf{2} = \begin{vmatrix} x_2 \\ y_2 \\ z_2 \end{vmatrix} \qquad (5.27)$$

The vector bar element is as follows:

$$\mathbf{B} = \begin{vmatrix} x_2 - x_1 \\ y_2 - y_1 \\ z_2 - z_1 \end{vmatrix} \tag{5.28}$$

The vector current is

$$\mathbf{c} = \begin{vmatrix} c_x \\ c_y \\ c_z \end{vmatrix} \tag{5.29}$$

Generally speaking, c_z is null.
The angle between current and B is

$$\cos(\alpha) = \frac{\mathbf{c} \cdot \mathbf{B}}{|\mathbf{c}||\mathbf{B}|} \tag{5.30}$$

5.4.3 Stiffness of the Normal Force

The normal force on B is

$$\mathbf{F} = |\mathbf{F}| \frac{\mathbf{B} \wedge (\mathbf{c} \wedge \mathbf{B})}{|\mathbf{B} \wedge (\mathbf{c} \wedge \mathbf{B})|} \tag{5.31}$$

That means that the x y and z components are:

$$\begin{aligned} \mathbf{F}_x &= |\mathbf{F}| \frac{\mathbf{E}_x}{|\mathbf{E}|} \\ \mathbf{F}_y &= |\mathbf{F}| \frac{\mathbf{E}_y}{|\mathbf{E}|} \\ \mathbf{F}_z &= |\mathbf{F}| \frac{\mathbf{E}_z}{|\mathbf{E}|} \end{aligned} \tag{5.32}$$

With:

$$\mathbf{E} = \mathbf{B} \wedge (\mathbf{c} \wedge \mathbf{B})$$

and

$$\mathbf{E} = \begin{vmatrix} E_x \\ E_y \\ E_z \end{vmatrix} \tag{5.33}$$

The x component of the derivative is

$$\mathbf{F}'_x = |\mathbf{F}|' \frac{\mathbf{E}_x}{|\mathbf{E}|} + |\mathbf{F}| \frac{\mathbf{E}'_x |\mathbf{E}| - \mathbf{E}_x |\mathbf{E}|'}{|\mathbf{E}|^2} \tag{5.34}$$

Which gives for the x y and z components:

$$\mathbf{F}'_x = |\mathbf{F}|'\frac{\mathbf{E}_x}{|\mathbf{E}|} + \frac{|\mathbf{F}|}{|\mathbf{E}|^2}\left\{\mathbf{E}'_x|\mathbf{E}| - \frac{\mathbf{E}_x}{|\mathbf{E}|}(\mathbf{E}_x\mathbf{E}'_x + \mathbf{E}_y\mathbf{E}'_y + \mathbf{E}_z\mathbf{E}'_z)\right\}$$

$$\mathbf{F}'_y = |\mathbf{F}|'\frac{\mathbf{E}_y}{|\mathbf{E}|} + \frac{|\mathbf{F}|}{|\mathbf{E}|^2}\left\{\mathbf{E}'_y|\mathbf{E}| - \frac{\mathbf{E}_y}{|\mathbf{E}|}(\mathbf{E}_x\mathbf{E}'_x + \mathbf{E}_y\mathbf{E}'_y + \mathbf{E}_z\mathbf{E}'_z)\right\} \qquad (5.35)$$

$$\mathbf{F}'_z = |\mathbf{F}|'\frac{\mathbf{E}_z}{|\mathbf{E}|} + \frac{|\mathbf{F}|}{|\mathbf{E}|^2}\left\{\mathbf{E}'_z|\mathbf{E}| - \frac{\mathbf{E}_z}{|\mathbf{E}|}(\mathbf{E}_x\mathbf{E}'_x + \mathbf{E}_y\mathbf{E}'_y + \mathbf{E}_z\mathbf{E}'_z)\right\}$$

For this assessment the derivative of $\mathbf{E}$ is required:

$$\mathbf{E}' = \mathbf{B}' \wedge (\mathbf{c} \wedge \mathbf{B}) + \mathbf{B} \wedge (\mathbf{c} \wedge \mathbf{B}') \qquad (5.36)$$

This leads to

$$\mathbf{E}' = 2(\mathbf{B}'.\mathbf{B})\mathbf{c} - (\mathbf{B}'.\mathbf{c})\mathbf{B} - (\mathbf{B}.\mathbf{c})\mathbf{B}' \qquad (5.37)$$

which is

$$\mathbf{E}'_x = 2(\mathbf{B}'.\mathbf{B})\mathbf{c}_x - (\mathbf{B}'.\mathbf{c})\mathbf{B}_x - (\mathbf{B}.\mathbf{c})\mathbf{B}'_x$$

$$\mathbf{E}'_y = 2(\mathbf{B}'.\mathbf{B})\mathbf{c}_y - (\mathbf{B}'.\mathbf{c})\mathbf{B}_y - (\mathbf{B}.\mathbf{c})\mathbf{B}'_y \qquad (5.38)$$

$$\mathbf{E}'_z = 2(\mathbf{B}'.\mathbf{B})\mathbf{c}_z - (\mathbf{B}'.\mathbf{c})\mathbf{B}_z - (\mathbf{B}.\mathbf{c})\mathbf{B}'_z$$

with

$$\mathbf{B}'.\mathbf{B} = \mathbf{B}_x\mathbf{B}'_x + \mathbf{B}_y\mathbf{B}'_y + \mathbf{B}_z\mathbf{B}'_z$$

$$\mathbf{B}'.\mathbf{c} = \mathbf{c}_x\mathbf{B}'_x + \mathbf{c}_y\mathbf{B}'_y + \mathbf{c}_z\mathbf{B}'_z \qquad (5.39)$$

$$\mathbf{B}.\mathbf{c} = \mathbf{B}_x\mathbf{c}_x + \mathbf{B}_y\mathbf{c}_y + \mathbf{B}_z\mathbf{c}_z$$

The derivative of the amplitude of the normal force is

$$|\mathbf{F}|' = \frac{1}{2}\rho C_d Dl_0|\mathbf{c}|^2 \left([\sin(\alpha)]^2\right)'$$

which is

$$|\mathbf{F}|' = \rho C_d Dl_0|\mathbf{c}|^2 \cos(\alpha)\sin(\alpha)\alpha' \qquad (5.40)$$

The derivative of α is

$$\alpha' = \frac{-1}{\sqrt{1 - (\frac{\mathbf{c}.\mathbf{B}}{|\mathbf{c}||\mathbf{B}|})^2}}\left[\frac{\mathbf{c}.\mathbf{B}}{|\mathbf{c}||\mathbf{B}|}\right]' \qquad (5.41)$$

That gives

$$\alpha' = \frac{-1}{\sqrt{1 - (\frac{\mathbf{c}.\mathbf{B}}{|\mathbf{c}||\mathbf{B}|})^2}}\left[\frac{\mathbf{c}}{|\mathbf{c}|}.\left(\frac{\mathbf{B}}{|\mathbf{B}|}\right)'\right] \qquad (5.42)$$

The derivative of the bar element direction is

$$\left(\frac{\mathbf{B}}{|\mathbf{B}|}\right)' = \frac{\mathbf{B}'|\mathbf{B}| - \mathbf{B}|\mathbf{B}|'}{|\mathbf{B}|^2} \tag{5.43}$$

That means that the derivative of α is

$$\alpha' = \frac{-1}{\sqrt{1 - (\frac{\mathbf{c}.\mathbf{B}}{|\mathbf{c}||\mathbf{B}|})^2}} \left(\frac{\mathbf{c}}{|\mathbf{c}|}\right) \cdot \left(\frac{\mathbf{B}'|\mathbf{B}| - \mathbf{B}|\mathbf{B}|'}{|\mathbf{B}|^2}\right) \tag{5.44}$$

or

$$\alpha' = \frac{-1}{|\mathbf{B}|^2|\mathbf{c}|\sin\alpha} \left\{ |\mathbf{B}| \left[c_x \mathbf{B}'_x + c_y \mathbf{B}'_y + c_z \mathbf{B}'_z \right] - (\mathbf{c}.\mathbf{B})|\mathbf{B}|' \right\} \tag{5.45}$$

In this case $\mathbf{B}'_x$ is the component along x of $\mathbf{B}'$.
The derivative of vector $\mathbf{B}$ is

$$\mathbf{B}' = \begin{vmatrix} \mathbf{B}'_x \\ \mathbf{B}'_y \\ \mathbf{B}'_z \end{vmatrix} \tag{5.46}$$

which is

$$\frac{\partial B_x}{\partial x_1} = \frac{\partial B_y}{\partial y_1} = \frac{\partial B_z}{\partial z_1} = -1$$
$$\frac{\partial B_x}{\partial x_2} = \frac{\partial B_y}{\partial y_2} = \frac{\partial B_z}{\partial z_2} = 1 \tag{5.47}$$

$$\frac{\partial B_x}{\partial y_1} = \frac{\partial B_x}{\partial y_2} = \frac{\partial B_x}{\partial z_1} = \frac{\partial B_x}{\partial z_2} = 0$$
$$\frac{\partial B_y}{\partial z_1} = \frac{\partial B_y}{\partial z_2} = \frac{\partial B_y}{\partial x_1} = \frac{\partial B_y}{\partial x_2} = 0 \tag{5.48}$$
$$\frac{\partial B_z}{\partial x_1} = \frac{\partial B_z}{\partial x_2} = \frac{\partial B_z}{\partial y_1} = \frac{\partial B_z}{\partial y_2} = 0$$

On vector form and for the nine coordinates of the triangular element it is

$$\frac{\partial \mathbf{B}}{\partial x_1} = \begin{vmatrix} -1 \\ 0 \\ 0 \end{vmatrix}$$

$$\frac{\partial \mathbf{B}}{\partial y_1} = \begin{vmatrix} 0 \\ -1 \\ 0 \end{vmatrix}$$

$$\frac{\partial \mathbf{B}}{\partial z_1} = \begin{vmatrix} 0 \\ 0 \\ -1 \end{vmatrix} \tag{5.49}$$

$$\frac{\partial \mathbf{B}}{\partial x_2} = \begin{vmatrix} 1 \\ 0 \\ 0 \end{vmatrix}$$

$$\frac{\partial \mathbf{B}}{\partial y_2} = \begin{vmatrix} 0 \\ 1 \\ 0 \end{vmatrix}$$

$$\frac{\partial \mathbf{B}}{\partial z_2} = \begin{vmatrix} 0 \\ 0 \\ 1 \end{vmatrix}$$

The derivative of the norm of vector $\mathbf{B}$ is

$$|\mathbf{B}|' = \frac{B_x B_x' + B_y B_y' + B_z B_z'}{|\mathbf{B}|} \tag{5.50}$$

Which gives for the nine coordinates of the triangular element:

$$\frac{\partial |\mathbf{B}|}{\partial x_1} = \frac{-B_x}{|\mathbf{B}|}$$
$$\frac{\partial |\mathbf{B}|}{\partial y_1} = \frac{-B_y}{|\mathbf{B}|}$$
$$\frac{\partial |\mathbf{B}|}{\partial z_1} = \frac{-B_z}{|\mathbf{B}|}$$
$$\frac{\partial |\mathbf{B}|}{\partial x_2} = \frac{B_x}{|\mathbf{B}|}$$
$$\frac{\partial |\mathbf{B}|}{\partial y_2} = \frac{B_y}{|\mathbf{B}|}$$
$$\frac{\partial |\mathbf{B}|}{\partial z_2} = \frac{B_z}{|\mathbf{B}|} \tag{5.51}$$

5.4.4 Stiffness of the Tangential Force

The tangential force on the bar element is

$$\mathbf{T} = |\mathbf{T}| \frac{\mathbf{F} \wedge (\mathbf{c} \wedge \mathbf{F})}{|\mathbf{F} \wedge (\mathbf{c} \wedge \mathbf{F})|} \tag{5.52}$$

Following the definition of $\mathbf{F}$:

$$\mathbf{T} = |\mathbf{T}| \frac{[\mathbf{B} \wedge (\mathbf{c} \wedge \mathbf{B})] \wedge \{\mathbf{c} \wedge [\mathbf{B} \wedge (\mathbf{c} \wedge \mathbf{B})]\}}{|[\mathbf{B} \wedge (\mathbf{c} \wedge \mathbf{B})] \wedge \{\mathbf{c} \wedge [\mathbf{B} \wedge (\mathbf{c} \wedge \mathbf{B})]\}|} \tag{5.53}$$

It follows:

$$\mathbf{T} = |\mathbf{T}| \frac{[(\mathbf{B}.\mathbf{B})(\mathbf{c}.\mathbf{c}) - (\mathbf{B}.\mathbf{c})^2](\mathbf{B}.\mathbf{c})\mathbf{B}}{|[(\mathbf{B}.\mathbf{B})(\mathbf{c}.\mathbf{c}) - (\mathbf{B}.\mathbf{c})^2](\mathbf{B}.\mathbf{c})\mathbf{B}|} \tag{5.54}$$

or

$$\mathbf{T} = |\mathbf{T}| \frac{[|\mathbf{B}|^2|\mathbf{c}|^2 - (|\mathbf{B}||\mathbf{c}|\cos\alpha)^2]|\mathbf{B}||\mathbf{c}|\cos\alpha\mathbf{B}}{|[|\mathbf{B}|^2|\mathbf{c}|^2 - (|\mathbf{B}||\mathbf{c}|\cos\alpha)^2]|\mathbf{B}||\mathbf{c}|\cos\alpha\mathbf{B}|} \tag{5.55}$$

and

$$\mathbf{T} = |\mathbf{T}| \frac{\cos\alpha\mathbf{B}}{|\cos\alpha||\mathbf{B}|} \tag{5.56}$$

The x y and z components are:

$$\begin{aligned}
\mathbf{T}_x &= |\mathbf{T}| \frac{\cos\alpha\mathbf{B}_x}{|\cos\alpha||\mathbf{B}|} \\
\mathbf{T}_y &= |\mathbf{T}| \frac{\cos\alpha\mathbf{B}_y}{|\cos\alpha||\mathbf{B}|} \\
\mathbf{T}_z &= |\mathbf{T}| \frac{\cos\alpha\mathbf{B}_z}{|\cos\alpha||\mathbf{B}|}
\end{aligned} \tag{5.57}$$

The derivative of $\mathbf{T}_x$ is:

$$\mathbf{T}'_x = |\mathbf{T}|' \frac{\cos\alpha\mathbf{B}_x}{|\cos\alpha||\mathbf{B}|} + |\mathbf{T}| \frac{(\cos\alpha\mathbf{B}_x)'|\cos\alpha||\mathbf{B}| - \cos\alpha\mathbf{B}_x(|\cos\alpha||\mathbf{B}|)'}{(|\cos\alpha||\mathbf{B}|)^2} \tag{5.58}$$

$$\begin{aligned}
\mathbf{T}'_x = &\,|\mathbf{T}|' \frac{\cos\alpha\mathbf{B}_x}{|\cos\alpha||\mathbf{B}|} \\
&+ \frac{|\mathbf{T}|}{|\cos\alpha||\mathbf{B}|}(\cos\alpha\mathbf{B}'_x - \sin\alpha\alpha'\mathbf{B}_x) \\
&- \frac{|\mathbf{T}|\cos\alpha\mathbf{B}_x}{(|\cos\alpha||\mathbf{B}|)^2}\left[|\cos\alpha|\frac{\mathbf{B}_x\mathbf{B}'_x + \mathbf{B}_y\mathbf{B}'_y + \mathbf{B}_z\mathbf{B}'_z}{|\mathbf{B}|} - \frac{\cos\alpha}{|\cos\alpha|}\sin\alpha\alpha'|\mathbf{B}|\right]
\end{aligned} \tag{5.59}$$

$$\begin{aligned}
\mathbf{T}'_x = &\,|\mathbf{T}|' \frac{\mathbf{T}_x}{|\mathbf{T}|} + \frac{|\mathbf{T}|}{|\cos\alpha||\mathbf{B}|}(\cos\alpha\mathbf{B}'_x - \sin\alpha\alpha'\mathbf{B}_x) \\
&- \frac{\mathbf{T}_x}{|\cos\alpha||\mathbf{B}|}\left[|\cos\alpha|\frac{\mathbf{B}_x\mathbf{B}'_x + \mathbf{B}_y\mathbf{B}'_y + \mathbf{B}_z\mathbf{B}'_z}{|\mathbf{B}|} - \frac{\cos\alpha}{|\cos\alpha|}\sin\alpha\alpha'|\mathbf{B}|\right]
\end{aligned} \tag{5.60}$$

$$\mathbf{T}'_y = |\mathbf{T}|' \frac{\mathbf{T}_y}{|\mathbf{T}|} + \frac{|\mathbf{T}|}{|\cos\alpha||\mathbf{B}|} (\cos\alpha\mathbf{B}'_y - \sin\alpha\alpha'\mathbf{B}_y)$$

$$- \frac{\mathbf{T}_y}{|\cos\alpha||\mathbf{B}|} \left[|\cos\alpha| \frac{\mathbf{B}_x}{\mathbf{B}_x}' + \mathbf{B}_y\mathbf{B}'_y + \mathbf{B}_z\mathbf{B}'_z|\mathbf{B}| - \frac{\cos\alpha}{|\cos\alpha|} \sin\alpha\alpha'|\mathbf{B}| \right] \quad (5.61)$$

$$\mathbf{T}'_z = |\mathbf{T}|' \frac{\mathbf{T}_z}{|\mathbf{T}|} + \frac{|\mathbf{T}|}{|\cos\alpha||\mathbf{B}|} (\cos\alpha\mathbf{B}'_z - \sin\alpha\alpha'\mathbf{B}_z)$$

$$- \frac{\mathbf{T}_z}{|\cos\alpha||\mathbf{B}|} \left[|\cos\alpha| \frac{\mathbf{B}_x\mathbf{B}'_x + \mathbf{B}_y\mathbf{B}'_y + \mathbf{B}_z\mathbf{B}'_z}{|\mathbf{B}|} - \frac{\cos\alpha}{|\cos\alpha|} \sin\alpha\alpha'|\mathbf{B}| \right] \quad (5.62)$$

The derivative of the amplitude of the tangential force is

$$|\mathbf{T}|' = f\frac{1}{2}\rho C_d Dl_0|\mathbf{c}|^2 \left([cos(\alpha)]^2 \right)' \frac{d}{2} \quad (5.63)$$

which is

$$|\mathbf{T}|' = -\frac{d}{2} f\rho C_d Dl_0|\mathbf{c}|^2 \cos(\alpha)\sin(\alpha)\alpha' \quad (5.64)$$

Chapter 6
The Node Element

Abstract The modelling for nodes in contact with the sea bed is described. The forces as well as the stiffness are given in case of contact as well as of friction.

Keywords Forces contact with the sea bed · Drag on the sea bed.

6.1 Principle

The contact of a marine structure with the sea bed has to be taken into account. It is of great importance for structures such as chains lying on the sea-bed or bottom trawls.

In the following sections a few forces related to this contact are described.

6.2 Contact on Bottom

In this model, the main hypothesis for these contact forces is that the bottom is elastic. That means that if a node is in contact with the bottom, the force reaction (N) is vertical and equal to the product of the node depth (m) in the soil by the soil stiffness (N/m).

6.2.1 Force Vector

The vertical force on a node due to its potential contact with the bottom is

$$if z < Z_b \quad F_z = B_k(Z_b - z) \tag{6.1}$$
$$if z \geq Z_b \quad F_z = 0 \tag{6.2}$$

D. Priour, *A Finite Element Method for Netting*, SpringerBriefs in Environmental Science, DOI: 10.1007/978-94-007-6844-4_6, © The Author(s) 2013

With:
F_z: the vertical force on the node (N),
B_k: the bottom stiffness (N/m),
Z_b: the vertical position of the bottom (m),
z: the vertical position of the node (m).

6.2.2 Stiffness Matrix

$$if z < Z_b \quad -\frac{\partial F_z}{\partial z} = B_k \tag{6.3}$$

$$if z \geq Z_b \quad -\frac{\partial F_z}{\partial z} = 0 \tag{6.4}$$

6.3 Drag on Bottom

Contact of a node with the bottom could lead to a wearing force. This force is taken into account when there is a movement of the structure on the bottom. This force is horizontal and opposite to the motion. This wearing depends on the depth on which the node digs the bottom, on the bottom stiffness, and on the node speed displacement on the bottom.

6.3.1 Force Vector

As mentioned earlier (Sect. 6.2, p. 87), the vertical force on a node due to its contact ($z < Z_b$) to the bottom is:

$$F_c = B_k(Z_b - z) \tag{6.5}$$

With:
F_c: the vertical force on the node due to the contact to the bottom (N),
B_k: the bottom stiffness (N/m),
Z_b: the vertical position of the bottom (m),
z: the vertical position of the node (m).

The drag force on the bottom has been modelled as a function of the displacement speed of the node on the bottom. Figure 6.1 shows this relation.

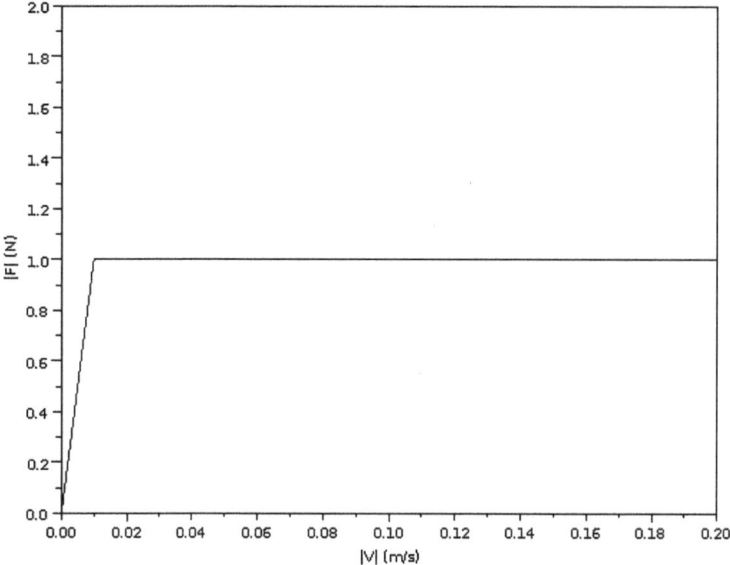

Fig. 6.1 Example of amplitude of wearing force |F| depending on the node displacement speed on the bottom |V|

$$if \, |\mathbf{V}| < V_l \ \ |\mathbf{F}| = F_c B_f \frac{|\mathbf{V}|}{V_l} \tag{6.6}$$

$$if \, |\mathbf{V}| \geq V_l \ \ |\mathbf{F}| = F_c B_f \tag{6.7}$$

With:

$$\mathbf{V} = \begin{vmatrix} V_x \\ V_y \\ V_z \end{vmatrix} \tag{6.8}$$

The components of speed are calculated as follows:

$$V_x = \frac{x - x_p}{\Delta t} \tag{6.9}$$

$$V_y = \frac{y - y_p}{\Delta t} \tag{6.10}$$

$$V_z = \frac{z - z_p}{\Delta t} \tag{6.11}$$

V_x (V_y, V_z): component of the speed of the node along the x (y, z) axis (m/s), x (y, z): coordinate of the node along the x (y, z) axis (m) calculated at time t,

x_p (y_p, z_p): previous coordinate of the node along the x (y, z) axis (m) calculated at time $t - \Delta t$.

Two cases are defined: a high-speed case ($|\mathbf{V}| \geq V_l$) and a low-speed case ($|\mathbf{V}| < V_l$). The wearing force is calculated in the two cases such as there is continuity between the two cases (at $|\mathbf{V}| = V_l$).

6.3.1.1 High-Speed

In this case, $|\mathbf{V}| \geq V_l$.

That means that the components of this force are the following:

$$F_x = -F_c B_f \frac{V_x}{|\mathbf{V}|} \tag{6.12}$$

$$F_y = -F_c B_f \frac{V_y}{|\mathbf{V}|} \tag{6.13}$$

$$F_z = -F_c B_f \frac{V_z}{|\mathbf{V}|} \tag{6.14}$$

6.3.1.2 Low-Speed

In this case, $|\mathbf{V}| < V_l$.

That means that the components of this force are the following:

$$F_x = -F_c B_f \frac{V_x}{V_l} \tag{6.15}$$

$$F_y = -F_c B_f \frac{V_y}{V_l} \tag{6.16}$$

$$F_z = -F_c B_f \frac{V_z}{V_l} \tag{6.17}$$

6.3.2 Stiffness Matrix

6.3.2.1 High-Speed

$$\frac{\partial F_x}{\partial x} = -\frac{F_c B_f}{|\mathbf{V}|^2} \frac{\partial V_x}{\partial x} \left[|\mathbf{V}| - \frac{V_x^2}{|\mathbf{V}|} \right] \tag{6.18}$$

$$\frac{\partial F_x}{\partial y} = -\frac{F_c B_f}{|\mathbf{V}|^2} \frac{\partial V_y}{\partial y} \left[-\frac{V_x V_y}{|\mathbf{V}|} \right] \tag{6.19}$$

$$\frac{\partial F_x}{\partial z} = B_k B_f \frac{V_x}{|\mathbf{V}|} - \frac{F_c B_f}{|\mathbf{V}|^2}\left[-\frac{V_x V_z}{|\mathbf{V}|}\frac{\partial V_z}{\partial z}\right] \tag{6.20}$$

$$\frac{\partial F_y}{\partial x} = \frac{F_c B_f}{|\mathbf{V}|^2}\left[\frac{V_x V_y}{|\mathbf{V}|}\frac{\partial V_x}{\partial x}\right] \tag{6.21}$$

$$\frac{\partial F_y}{\partial y} = -\frac{F_c B_f}{|\mathbf{V}|^2}\frac{\partial V_y}{\partial y}\left[|\mathbf{V}| - \frac{V_y^2}{|\mathbf{V}|}\right] \tag{6.22}$$

$$\frac{\partial F_y}{\partial z} = B_k B_f \frac{V_y}{|\mathbf{V}|} - \frac{F_c B_f}{|\mathbf{V}|^2}\left[-\frac{V_x V_z}{|\mathbf{V}|}\frac{\partial V_z}{\partial z}\right] \tag{6.23}$$

$$\frac{\partial F_z}{\partial x} = \frac{F_c B_f}{|\mathbf{V}|^2}\left[\frac{V_x V_z}{|\mathbf{V}|}\frac{\partial V_x}{\partial x}\right] \tag{6.24}$$

$$\frac{\partial F_z}{\partial y} = \frac{F_c B_f}{|\mathbf{V}|^2}\left[\frac{V_y V_z}{|\mathbf{V}|}\frac{\partial V_y}{\partial y}\right] \tag{6.25}$$

$$\frac{\partial F_z}{\partial z} = B_k B_f \frac{V_z}{|\mathbf{V}|} - \frac{F_c B_f}{|\mathbf{V}|^2}\left[\frac{\partial V_z}{\partial z}|\mathbf{V}| - \frac{V_z^2}{|\mathbf{V}|}\frac{\partial V_z}{\partial z}\right] \tag{6.26}$$

With:

$$\frac{\partial V_x}{\partial x} = \frac{1}{\Delta t} \tag{6.27}$$

$$\frac{\partial V_y}{\partial y} = \frac{1}{\Delta t} \tag{6.28}$$

$$\frac{\partial V_z}{\partial z} = \frac{1}{\Delta t} \tag{6.29}$$

The stiffness matrix becomes:

$$K = -\frac{B_f F_c}{|\mathbf{V}|^2 \Delta t}\begin{pmatrix} \frac{V_x^2}{|\mathbf{V}|} - |\mathbf{V}| & \frac{V_x V_y}{|\mathbf{V}|} & \frac{V_x V_z}{|\mathbf{V}|} \\ \frac{V_x V_y}{|\mathbf{V}|} & \frac{V_y^2}{|\mathbf{V}|} - |\mathbf{V}| & \frac{V_y V_z}{|\mathbf{V}|} \\ \frac{V_x V_z}{|\mathbf{V}|} & \frac{V_y V_z}{|\mathbf{V}|} & \frac{V_z^2}{|\mathbf{V}|} - |\mathbf{V}| \end{pmatrix} - \frac{B_f B_k}{|\mathbf{V}|}\begin{pmatrix} 0 & 0 & V_x \\ 0 & 0 & V_y \\ 0 & 0 & V_z \end{pmatrix} \tag{6.30}$$

6.3.2.2 Low-Speed

$$\frac{\partial F_x}{\partial x} = -\frac{F_c B_f}{V_l}\frac{\partial V_x}{\partial x} \tag{6.31}$$

$$\frac{\partial F_x}{\partial y} = 0 \tag{6.32}$$

$$\frac{\partial F_x}{\partial z} = B_k B_f \frac{V_x}{V_l} \tag{6.33}$$

$$\frac{\partial F_y}{\partial x} = 0 \tag{6.34}$$

$$\frac{\partial F_y}{\partial y} = -\frac{F_c B_f}{V_l} \frac{\partial V_y}{\partial y} \tag{6.35}$$

$$\frac{\partial F_y}{\partial z} = B_k B_f \frac{V_y}{V_l} \tag{6.36}$$

$$\frac{\partial F_z}{\partial x} = 0 \tag{6.37}$$

$$\frac{\partial F_z}{\partial y} = 0 \tag{6.38}$$

$$\frac{\partial F_z}{\partial z} = B_k B_f \frac{V_z}{V_l} - \frac{F_c B_f}{V_l} \frac{\partial V_z}{\partial z} \tag{6.39}$$

The stiffness matrix becomes:

$$K = -\frac{B_f}{V_l} \begin{pmatrix} \frac{F_c}{\Delta t} & 0 & -B_k V_x \\ 0 & \frac{F_c}{\Delta t} & -B_k V_y \\ 0 & 0 & \frac{F_c}{\Delta t} - B_k V_z \end{pmatrix} \tag{6.40}$$

Chapter 7
Validation

Abstract Some cases of validation of the modelling are given. These are compared with flume tank tests, sea trials, and other models. They are about tractrix shape of netting, netting stretched under its own weight, netting of hexagonal meshes deformed by a water current, hydrostatic pressure in a bag of netting, cod-end of trawl with a catch, bottom trawl at sea, cubic fish farm and bending of a rigid cable.

Keywords Tractrix shape · Stretched netting · Deformation of hexagonal meshes · Hydrostatic pressure on netting · Cod-end of trawl · Bottom trawl · Cubic fish farm · Bending cable

7.1 Tractrix

The shape of the meridian of a cylinder of netting of inextensible twines held between two circular rings is a tractrix, if the axis of the two rings are identical and coplanar with meshes diagonals.

In the case of a cylinder of stretched netting of 100 meshes around, 50 meshes along, a radius of 1 m at one extremity and 0.048599 m at the other, and a mesh side of 0.05 m, the shape is as displayed in Fig. 7.1 [14].

The accuracy of the model depends on the number of nodes used (Table 7.1). The model uses 32–662 nodes and two planes of symmetry.

7.2 Diamond Mesh Netting Stretched by its Weight

This check is done by comparing the results of the model based on triangular elements with a model where each twine is modelled by an elastic bar. This comparison is taken from [16].

D. Priour, *A Finite Element Method for Netting*, SpringerBriefs in Environmental Science, DOI: 10.1007/978-94-007-6844-4_7, © The Author(s) 2013

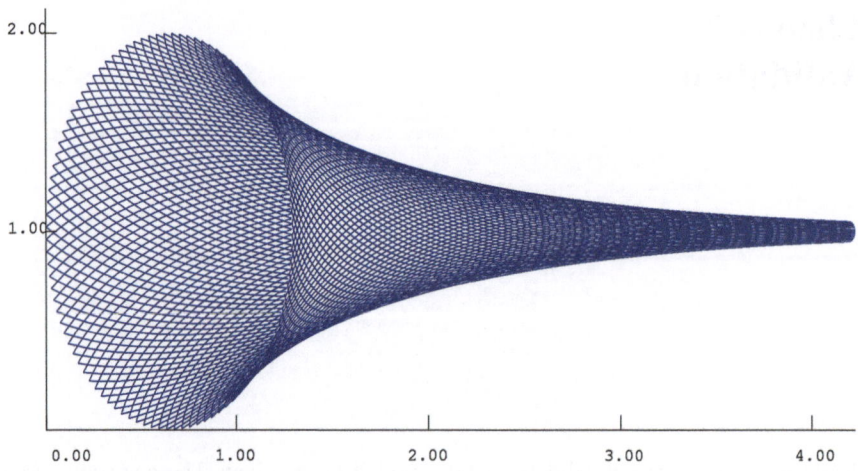

Fig. 7.1 Cylinder of inextensible netting held between two circular rings

Table 7.1 Tractrix shape (x, y) and accuracy (%) of the model, where x and y are the analytical solution; x is along the axis and y is radial. The accuracy on y depends on the number of nodes in the model (from 32 to 662)

x (m)	y (m)	662 (%)	298 (%)	84 (%)	32 (%)
0	1				
0.403501	0.739032	0.02	0.22	1.4	−1.1
0.844094	0.546168	0.00	0.19	1.2	−2.7
1.303628	0.403636	−0.01	0.14	1.0	−1.8
1.773173	0.2983	0.00	0.19	1.5	−2.3
2.248093	0.220453	−0.02	0.17	1.3	−2.4
2.725923	0.162922	−0.03	0.15	1.0	−3.6
3.205334	0.120404	−0.07	0.18	1.0	−2.8
3.685607	0.088983	−0.11	0.17	0.6	−3.2
4.166349	0.065761	−0.15	0.16	0.2	−1.9
4.647348	0.048599				

The mesh panel is square and consists of 1600 meshes. The elongation rigidity (*EA*) of the twines is 10000 N, their diameter is 0.01 m, the side of the mesh is 1.2 m, the length of the upper edge is 32 m, and the density of the net is 2000 kg/m^3 (Fig. 7.2).

The model uses 1050 triangular elements and 512 nodes with a vertical plane of symmetry (Figs. 7.2 and 7.3b). The comparison is made with a reference model where each side of mesh (twine vector) is modelled with an elastic bar (Fig. 7.3a). This reference model uses 3136 bars and 1625 nodes with a plane of symmetry. The forms calculated by the two models are quite similar (Fig. 7.3).

The forces involved here are the netting weight and the twine tension (Sects. 4.3.9 p. 68 and 4.3 p. 36).

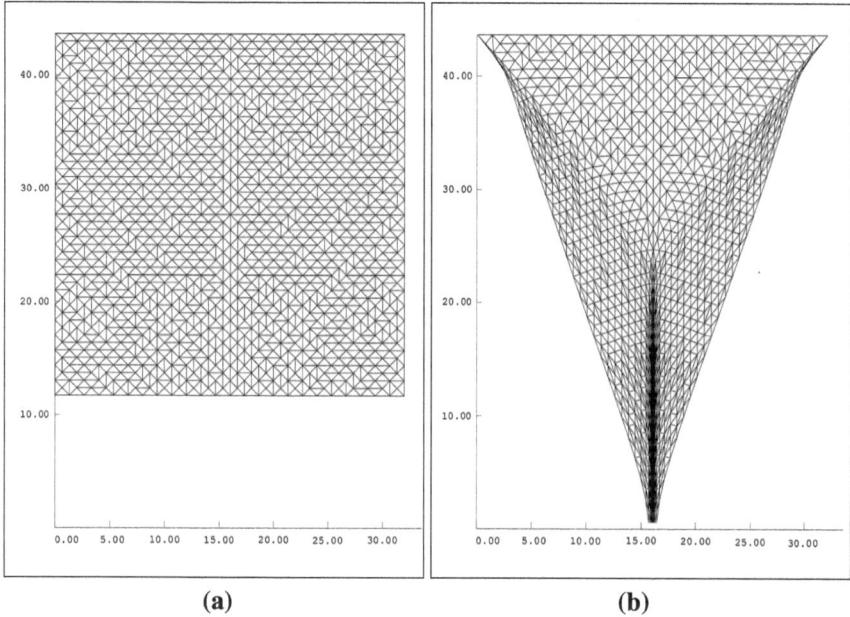

Fig. 7.2 Calculation of the shape of a net held by its *top* border. The initial shape of the model is unbalanced (**a**) and the final one is balanced (**b**). Only the triangular elements are represented

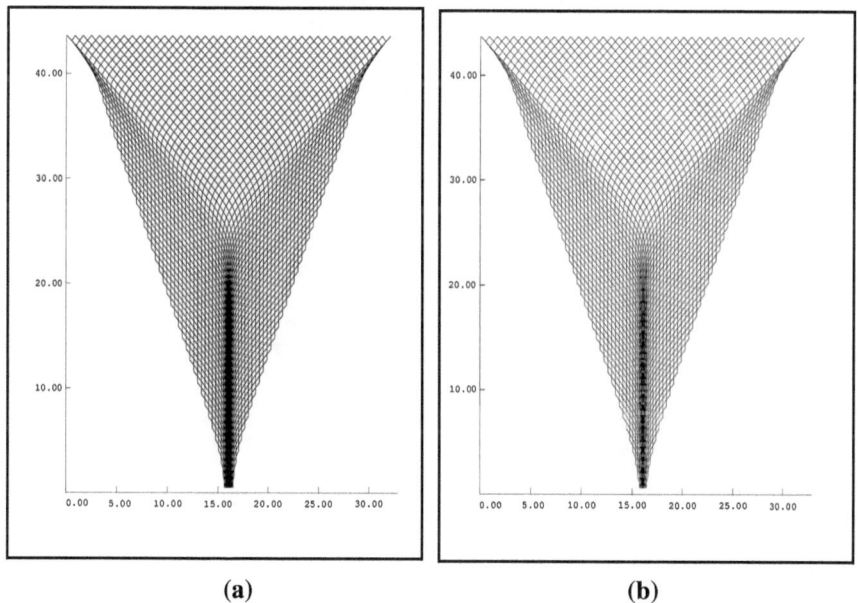

Fig. 7.3 Equilibrium of a net held by its top edge and stretched by its own weight: **a** model where each twine is modelled as an elastic bar; **b** model using triangular elements, with only the twines drawn

7.3 Hexagonal Mesh Net Held Vertically in the Current

The results of the model using triangular elements for netting with hexagonal meshes
are compared with those of a model using bar elements for each twine. The mesh
panel is square and consists of 18 by 33 meshes and 3564 twines. The elongation
rigidity of the twines is 3000 and 0.0003 N in compression. The diameter of the twines
is 1 mm, and their length is 19 mm. The length of each edge is 1 m. The density of
the material is considered equal to that of sea water (1025 kg/m^3). The net is held
by its four edges perpendicular to a current of 1 m/s of sea water.

The first model uses 924 triangular elements and 495 nodes (Fig. 7.4a, b), whereas
the second model uses 3564 bars and 2446 nodes (Fig. 7.4c).

The results of the two models are similar. The maximum displacement is 0.182 m
for the first model and 0.184 m for the second. The drag force is 54.10 N for the first
and 54.04 N for the second.

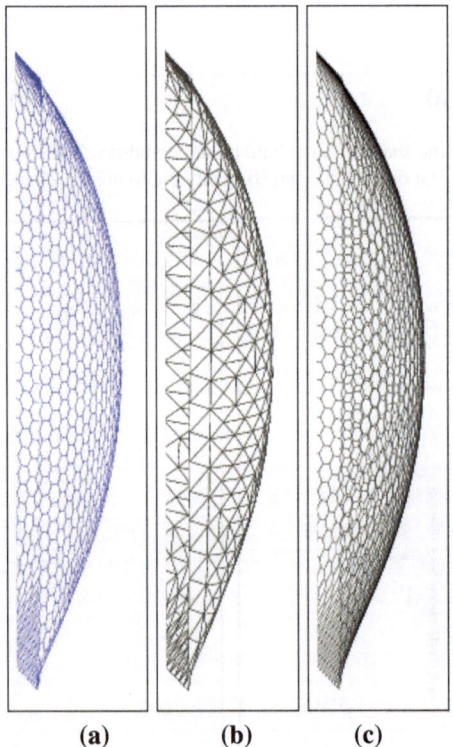

(a)   **(b)**   **(c)**

Fig. 7.4 Equilibrium of a net held by its four edges in a current perpendicular: **a** the twines in
the model using triangular elements; **b** the triangular elements; **c** the twines in the model using bar
elements. The shapes are similar

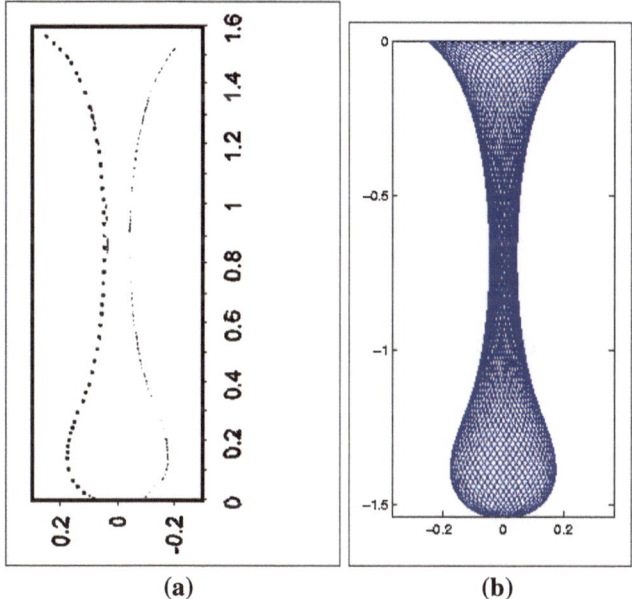

Fig. 7.5 Bag of netting with 26.5 kg of water. Comparison between measurements (**a**) and the model using triangular elements (**b**). Only twines are shown in (**b**)

Convergence is obtained in 29 iterations with the first model compared with 296 iterations for the second model. This acceleration is related to the reduction in the number of nodes in the model using triangular elements.

This comparison is based on [17] .

7.4 Hydrostatic Pressure

The results of the model using triangular elements are compared with measurements made by [13]. These measures involve a net bag partially filled with water bags (Fig. 7.5). The pressure from the weight of the bags is implemented as in Sect. 4.3.6 (p. 64), but in this case the pressure is modelled as a hydrostatic pressure:

$$p = \rho g h \tag{7.1}$$

p: pressure exerted by the catch on the net (Pa),
ρ : density of water (kg/m^3),
g: gravity (9.81 m/s^2),
h: height in relation to the upper limit of the catch (m).
The test conditions are as follows:

Mesh size: 37.2 mm,
Number of meshes around: 50,
Number of meshes along: 50,
Catch volume: $0.0265\,\text{m}^3$,
Catch density (ρ): $1000\,\text{kg/m}^3$,
Radius of the hoop above: 0.25 m

The model uses 742 nodes, 1360 triangular elements, one bar for closing the netting at the bottom, and two symmetry planes. This comparison comes from [18].

7.5 Cod-End with Catch in the Current

A cod-end is the backmost part of a trawl where the catch of fish builds up. The results of the model are compared with measurements made in test tank on cod-ends partially filled with water [1]. The pressure of the catch is implemented here as follows (see Sect. 4.3.6, p. 64):

$$p = \frac{1}{2}\rho C_d v^2 \tag{7.2}$$

p: catch pressure on the net (Pa),
ρ : density of water (kg/m^3),
C_d: drag coefficient (1.4),
v: current amplitude (m/s).

The distance between the front of the catch and the extremity of the cod-end is inserted into the model as data because this distance was measured during the tests. Figure 7.6 shows the model output (net) and the flume tank measurements (cross). The comparison shows that the model gives a pretty good description of the cod-end with the catch.

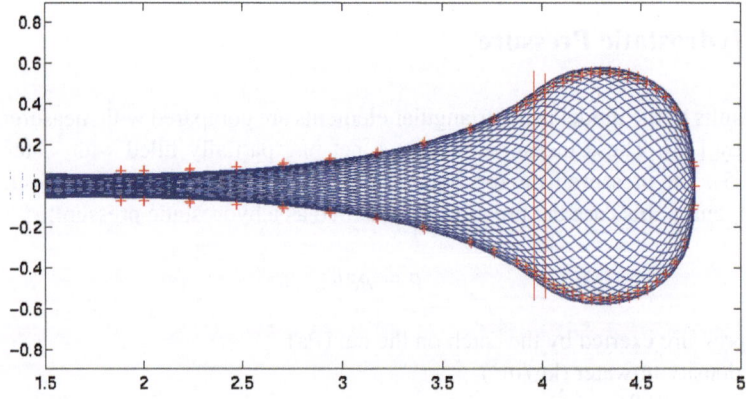

Fig. 7.6 Comparison of flume tank tests (*cross*) and the numerical model outputs (*mesh*) for a scale (1/3) model of North Sea cod-end with 300 kg of catch

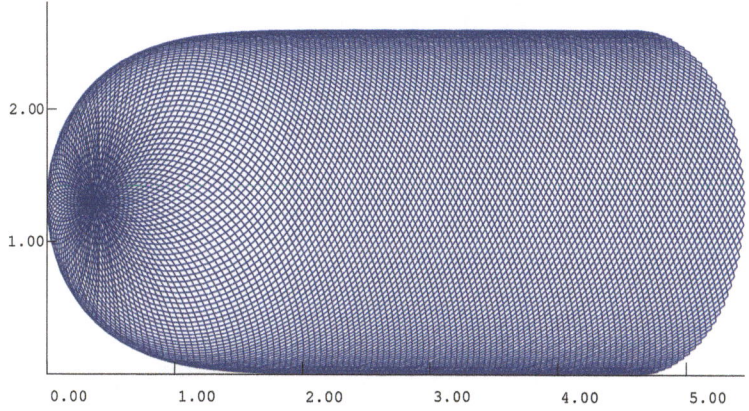

Fig. 7.7 Cod-end of netting subject to constant internal pressure

7.6 Full Cod-End

A long and full cod-end subject to constant internal pressure presents a maximal diameter. This maximal diameter depends on the number of meshes around N and the mesh side m by the following analytical equation [14]:

$$D_{max} = 4\frac{Nm}{\pi\sqrt{6}}$$

In the case of a cod-end close at one extremity of 100 meshes along (N), and a mesh side of 0.05m (m), the shape is as displayed in Fig. 7.7. The accuracy of the model on the maximal diameter is 0.015%.

7.7 Bottom Trawl

Several series of measurements on a bottom trawl were carried out during a sea trial on a French vessel. The results of the numerical model were compared with these measurements [20] (Fig. 7.8, Table 7.2).

For the measurements at sea, the vessel was equipped with measurement systems suitable for trawling. Several measurements were carried out:

- the position of the doors (immersion and distance),
- the distance between the headline and the bottom,
- the speed over ground and speed relative to the water,
- the warps and bridles tension.

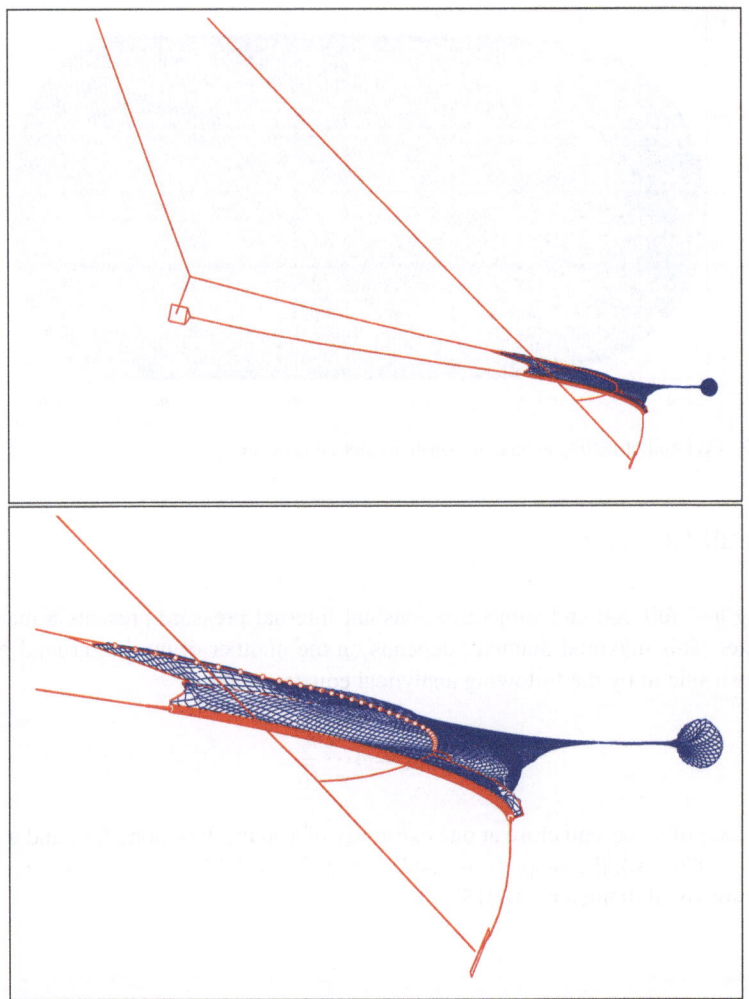

Fig. 7.8 Shape of the *bottom* trawl assessed by the model. Only 1 twine on 5 is drawn. The doors are modelled by 4 bars. The catch is on the *right*

Table 7.2 Differences between tests at sea and simulation

	Mean − SD	Mean + SD	Simulation
Warp tension (kg)	1966	3121	2300
Top bridle tension (kg)	864	1370	980
Bottom bridle tension (kg)	609	972	830
Vertical opening (m)	3.5	4.3	3.4

SD: standard deviation

For the modelling of such complex structure, numerous parameters have to be determined:

- the design of the netting, which includes the number of netting panels, the size of each panel in meshes number, the mesh size, the twine diameter, stiffness, density and drag coefficients.
- the design of cables, the number of cables, the cable stiffness, length, density and drag coefficients,
- the links between elements, which define among other the seams between netting panels,
- the modelling of doors, which is determined in the present case by four bars (visible on Fig. 7.8), on which the weight, the drag and the lift are applied. These forces are calculated from data provided by the door maker,
- the water depth and the towing speed,
- the catch volume and the drag coefficient on the catch,
- the wearing coefficient on the sea bottom and the stiffness of the bottom,
- the floats repartition, which are on the headline in the present case, with the floatability and volume of each float.

Measurements on the trawl are highly variable. The results of model calculation are generally close to measured quantities.

7.8 Cubic Fish Cage

Tests were carried out on models of a fish cage in the flume tank of Boulogne/mer [21]. The cage consisted of 4 side panels of 23 horizontal by 26 vertical meshes and a bottom panel of 23 by 23 meshes. The net had a mesh side of 35 mm and a twine diameter of 2.2 mm. The four bottom corners were tightened with 3 kg of lead sinkers. The size of the cage top was 1 by 1 m. The water speed was 0.5 m/s. Figure 7.9 compares the flume tank test and the simulation.

(a)   **(b)**

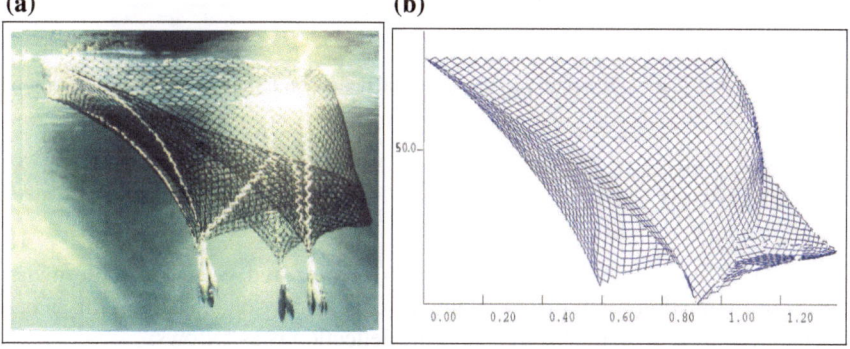

Fig. 7.9 Qualitative comparison between the deformation of a cubic cage in a flume tank (**a**) and simulation (**b**)

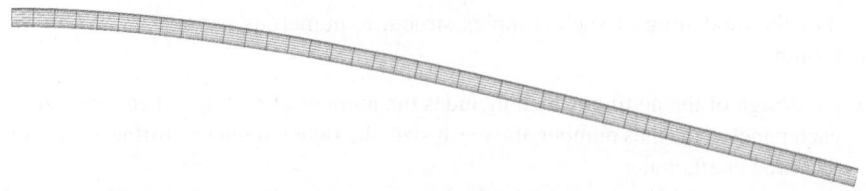

Fig. 7.10 Vertical deflection of a beam calculated with the model. The beam is fixed on the *left* and free to bend on its own weight on the *right*. The conditions are the same as in the text except for the bending rigidity, which is ($EI = 16.493$ N.m^2), ten times less than the case of Table 7.3 and Fig. 7.11 to highlight the deformation

Table 7.3 Vertical deflection of the beam deflection calculated with the model in function of bar elements number and error relative to the analytical deflection (18.2 mm)

Number of bars	5	8	10	12	16	20	30	40
Simulated deflection (mm)	18.9	18.5	18.4	18.3	18.3	18.3	18.2	18.2
Error %	4.0	1.5	0.97	0.67	0.36	0.23	0.082	0.039

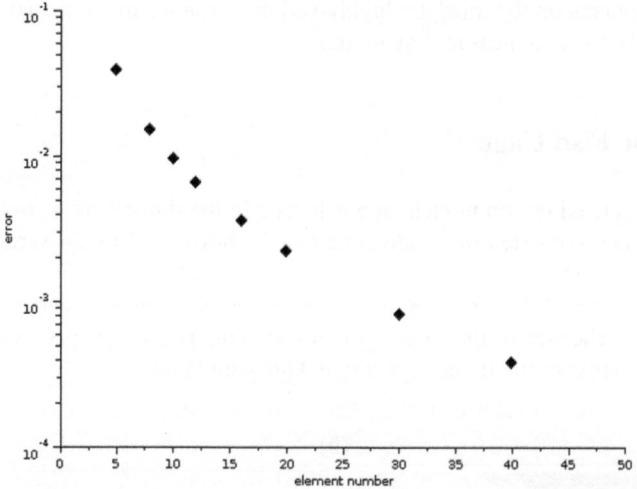

Fig. 7.11 Error of the model relative to the analytical deflection in function of the number of bar elements

7.9 Bending of Cable

The model of bending of cables (Sect. 5.3, p. 74) is compared with a beam deformation (Fig. 7.10) in the thin beam theory. In this case the deflection is well known. In case of a cantilever the analytical equation of the deflection is as follows:

$$y = \frac{-Wl^4}{8EI} \tag{7.3}$$

y: the vertical deflection of the free extremity of the cantilever (m),
l: the length of the cantilever (m),
w: the linear weight of the cantilever (N/m),
EI: the bending rigidity (N.m^2).

In case of a beam 1 m long (l), with a density of iron (7800 kg/m^3), a diameter of 2 cm, and a rigidity (EI) of 164.93 N.m^2, the deflection is 18.2 mm.

Table 7.3 and Fig. 7.11 show the vertical deflection of the beam calculated with the model in function of bar element number. The model is shown to be valid. The larger the number of bar elements, the smaller the error.

References

1. Anon (1999) PREMECS FAIR Program CT96 1555, Final report 1t December 1996 31st November 1999
2. Bessonneau JS, Marichal D (1998) Study of the dynamics of submerged supple nets. Ocean Eng 27(7):563–583
3. Chang SY (2004) Studies of Newmark method for solving non linear systems: (I) Basic analysis. J Chin Inst Eng 27(5):651–662
4. Deuflhard P (2004) Newton methods for non-linear problems, Affine invariance and adaptive algorithms. Springer series in computational mathematics. ISSN 0 179–3632. ISBN 3-540-21099-7
5. Desai CS, Abel JF (1972) Introduction to the finite element method: a numerical method for engineering analysis. Van Nostrand Reinhold, New York
6. Ferro RST (1988) Computer simulation of trawl gear shape and loading.In: Proceedings of word symposium on fishing gear and fishing vessel design.Marine Institute, Saint Johns, pp 259–262
7. Hallam MG, Heaf NJ, Wootton LR (1977) Dynamics of marine structures. CIRIA Underwater Engineering Group, Londres
8. Landweber L, Protter MH (1947) The shape and tension of a lightflexible cable in a uniform current. J Appl Mech 14:121–126
9. Le Dret H, Priour D, Lewandowski R, Chagneau F (2004) Numerical simulation of a cod end net. Part 1. Equilibrium in a uniform flow. J Elast 76(2):139–162
10. Lee C-W, Lee J-H, Cha B-J, Kim H-Y, Lee J-H (2005) Physical modeling for underwater flexible systems dynamic simulation. Ocean Eng 32:331–347
11. Niedzwiedz G, Hopp M (1998) Rope and net calculations applied to problems in marine engineering and fisheries research. Arch Fish Mar Res 46:125–138
12. O'Neill FG (2004) The influence of bending stiffness on thedeformation of axi-symmetric networks.In: Proceedings of OMAE'04, June 20–25, 2004. Vancouver Canada
13. O'Neill FG, O'Donoghue T (1997) The fluid dynamic loadingon catch and the geometry of trawl cod-ends.Proc Roy Soc London Ser Math Phys Sci 453:1631–1648
14. O'Neill FG, Priour D (2009) Comparison and validation of two models of netting deformation. J Appl Mech 76(5):1–7
15. O'Neill FG, Xu L (1994) Twine flexural rigidity and meshresistance to opening, ICES CM/B:31
16. Priour D (1999) Calculation of net shapes by the finite element method with triangular elements. Commun Numer Meth Eng 15:755–763
17. Priour D (2002) Analysis of nets with hexagonal mesh using triangular elements. Int J Numer Meth Eng 56:1721–1733. doi:10.1002/nme.635

18. Priour D (2005) FEM modelling of flexible structures made of cables,bars and nets. In: Soares G, Garbatov Y, Fonseca N (eds) Proceedings of the IMAM conference:Maritime transportation and exploitation of ocean andcoastal resources. Taylor and Francis, London, pp 1285–1292

19. Priour D (2006) Twines equilibrium in a finite element dedicatedto hexagonal mesh netting.In: ESAIM: Proceedings, October 2007, vol. 22, pp 140–149

20. Priour D (2012) Rapport final du projet EFFICHALUT,Rapport Ifremer/DCB/RDT/HO/R12-001

21. Repecaud M, Rodier P (1993) Note interne IFREMER,Compte rendu dessais: Cages pour l'elevage du poisson en mer, DITI/NPA/93.020

22. Richtmeyer RD (1941) Design and operation of mark IV magnetic mine sweeping gear.Bureau of ships scientific group report No12. January 1941

23. Rivlin RS (1955) Plane strain of a net formed by inextensiblecords. Indiana Univ Math J 4:951–974

24. Tsukrov I, Eroshkin O, Fredriksson D, Swift MR, Celikkol B (2003)Finite element modeling of net panels using a consistent net element.Ocean Eng 30(2):251–270

25. Zienkiewicz OC, Taylor RL (1989), The finite element method, McGraw-Hill Book Company

Index

D. Priour, *A Finite Element Method for Netting*, SpringerBriefs in Environmental Science, DOI: 10.1007/978-94-007-6844-4, © The Author(s) 2013